Carbon Neutrality in the Agri-food Sector

Alessia Acampora • Luca Ruini
Carlo Alberto Pratesi • Maria Claudia Lucchetti

Carbon Neutrality in the Agri-food Sector

Challenges and Opportunities in a Dynamic Setting

Springer

Alessia Acampora
Department of Business Studies
Roma Tre University
Rome, Italy

Luca Ruini
Global Supply Chain - People Safety,
Environment & Energy Vice President - Barilla
G. e R. Fratelli S.p.A
Parma, Italy

Carlo Alberto Pratesi
Department of Business Studies
Roma Tre University
Rome, Italy

Maria Claudia Lucchetti
Department of Business Studies
Roma Tre University
Rome, Italy

ISBN 978-3-030-88047-7 ISBN 978-3-030-88048-4 (eBook)
https://doi.org/10.1007/978-3-030-88048-4

This Springer imprint is published by the registered company Springer Nature Switzerland AG
The registered company address is: Gewerbestrasse 11, 6330 Cham, Switzerland

Preface

This book deals with the in-depth study of sustainability issues in the agri-food sector. In particular, a critical analysis of the current situation was developed and the future prospects of the sector on the issue of managing the environmental variable and the impacts relating to food production and consumption have been analyzed. Furthermore, the state of implementation and best practices relating to the carbon neutrality model in the agri-food sector were deepened and models for the development of a new food production system were proposed. The main objectives of the book are to analyze the current situation and trends regarding carbon neutrality schemes and the connection with other greening programs; to identify and analyze the carbon-related labels, their methodology, and their conformity assessment mechanisms; and to understand possible key drivers for carbon neutrality or low carbon strategies in the agri-food sector.

Particularly, the goal of this book is to study how to boost interest and help companies achieve their carbon neutrality and low carbon goals and to provide academics, companies, and policymakers with a useful guide to implement and explore carbon neutrality pathways in the agri-food sector. In the book it emerges clearly that citizens and consumers are demanding solutions for climate change and companies are acting in response to these demands and are developing carbon-neutral strategies to promote the transition towards a low-carbon society. This tool is easy to communicate, and companies are using it as marketing leverage. The book focuses on the agri-food sector as agriculture is simultaneously the main cause and the main victim of the ongoing environmental crisis. Agricultural practices, in fact, produce significant volumes of GHG emissions, the main cause of climate change. Nevertheless, this sector suffers the most negative impacts of climate change, in terms of reduced productivity and increased risks related to food safety. Experts argue that even low levels of global warming will reduce agricultural production and trigger greater variability in production in lower-latitude regions of the world. However, agriculture has a great potential in the fight against climate change, it is the only sector that has the capacity to remove GHGs safely and cost effectively from the atmosphere without reducing productivity. In this context, agri-food companies are implementing carbon-neutral strategies for their brands and products. A lot of case studies have been presented in this book showing how different are the approaches that companies are adopting. There is not a magic formula to reach the carbon neutrality that is universally valid. Companies are combining the tools available on the market according to their different goals, and their specificity. Companies are also experimenting with new approaches, acting in a void of specific definitions and governance. This book presents an updated guide to carbon neutrality exploring all these different steps, tools, and approaches.

As this book shows, the frontrunners of this transition are surely the companies of the agri-food sector. These companies have identified a specific competitive advantage and are exploiting it to become the pioneers and the reference model for the carbon-neutral implementation. Indeed, developing a carbon-neutral strategy can help companies set long-term strategies to reduce their carbon footprints and boost the transition towards a low-carbon agri-food sector. It represents a long-term umbrella project that can help companies implement short-term solutions while developing long-term carbon reduction projects. However, in this book, we also pointed out the carbon-neutrality path is far from perfect. Indeed, the recent interest in this process triggered the need to review this instrument in terms of governance and robustness. In particular, the carbon-neutrality process should get out of the initial development phase and establish clearer rules for good governance if it wants to establish itself as a functional tool for the low-carbon transition. The book analyzes the different aspects that characterize the carbon-neutrality strategy in the agri-food sector also highlighting shortcomings and difficulties specifically related to different products. From the analysis it emerged that the governance system of carbon neutrality still presents certain drawbacks that need to be overcome to assure the correct functioning of the system.

Rome, Italy Alessia Acampora

Parma, Italy Luca Ruini

Rome, Italy Carlo Alberto Pratesi

Rome, Italy Maria Claudia Lucchetti

Acknowledgments

The authors would like to acknowledge the support of the Food and Agriculture Organization of the United Nations (FAO) for providing funds that supported this work.

Indeed, this book presents the results of a research work the authors developed for the Food and Agriculture Organization of the United Nations (FAO) and the European Bank for Reconstruction and Development (EBRD). The book also benefits from the precious suggestions and reviews of FAO colleagues Nuno Santos, Jacopo Monzini, and Maria Ricci.

Contents

1 Carbon Neutrality: If Not Now, When? 1
- 1.1 Citizens' Perception of Climate Change Seriousness 1
- 1.2 Climate Change and Carbon Neutrality Policies 4
  - 1.2.1 Long-Term Low Greenhouse Gas Emission Development Strategies and Nationally Determined Contributions (NDCs) 8
  - 1.2.2 The UNFCCC's Measurement, Reporting and Verification (MRV) Framework and Enhanced Transparency Framework (ETF) 13
- 1.3 Countries' Carbon Neutral Policies 17
- 1.4 Carbon Neutrality Definitions 18
- References 24

2 Carbon Neutrality in the Agri-Food Sector 29
- 2.1 Agri-Food Systems Are Facing a Growing Climate Change Problem and Opportunity 29
- 2.2 Agriculture as Source of Solutions in Fighting Climate Change: Turning a Planetary Pollutant into an Asset 31
- 2.3 Mitigation Projects in the Agri-Food Sector 42
  - 2.3.1 Tree-Planting Initiatives 44
  - 2.3.2 Offset Projects in the Agri-Food Sector 45
- References 47

3 Putting a Farm into an Emissions Test Lab 51
- 3.1 Life Cycle Assessments (LCA) and Carbon Footprints 51
- 3.2 Carbon Footprint (CF) 51
  - 3.2.1 Organizational Carbon Footprint (OCF) 52
  - 3.2.2 Product Carbon Footprint (PCF) 53
- 3.3 The GHG Protocol Initiative 54
  - 3.3.1 The GHG Protocol for Guidance on Carbon Removals and Land Use 54

3.3.2 Scope of the GHG Protocol Corporate Standard 57
3.4 Carbon Footprint in Agri-Food Sector: Methodological Issues 58
References 62

4 Governance Issues in Carbon Emissions: Can a Food Emissions Gate Be Avoided? 63
4.1 'Emissions Gate' and Increasing Concerns Over Greenwashing . . . 63
4.1.1 Greenwashing and Companies' Sustainability Reporting . . 65
4.1.2 Sustainability Reporting: What Investors Are Asking 67
4.2 Carbon Disclosure Initiatives 68
References 70

5 The Governance of the Carbon Neutrality System: Is Carbon Neutrality a Tool for the Low-Carbon Transition? 71
5.1 Carbon Neutrality: A Nudging Strategy Towards the Decarbonization 71
5.2 Different Approaches to Use the Carbon Footprint 72
5.2.1 Carbon quantification /Measurement 73
5.2.2 Carbon Reduction 73
5.2.3 Carbon Neutrality 74
5.2.4 Climate/Carbon Positive 74
5.3 Carbon Offsets 76
5.3.1 Carbon Offset Standards 77
5.3.2 Voluntary Offset Market Trends 81
5.4 Carbon Insetting 82
5.4.1 The Plan Vivo Standard for Insetting 86
5.4.2 Insetting Case Studies 87
5.5 Targets and Scopes of the Companies' Carbon Neutrality Path . . . 88
5.5.1 Targets of the Carbon Neutrality 88
5.5.2 Scope of the Carbon Neutrality 93
5.6 Voluntary Standards and Carbon Neutrality 96
5.6.1 The PAS 2060: 2014 Specification for the Demonstration of Carbon Neutrality 99
5.6.2 The Climate Active Carbon Neutral Standard 102
5.7 Verification, Certification, Accreditation and Labels 103
5.8 Carbon Neutral Labels 107
References 113

6 How Difficult Is It to Reach Carbon Neutrality? Tales of Different Agri-Food Chains 115
6.1 The Role of Different Actors of the Agri-Food Supply Chain in Achieving Sector' Carbon Neutrality 115
6.2 Differentiated Impacts for Different Food Products 121
6.3 Achieving Carbon Neutrality in Different Agri-Food Chains: Case Studies Analysis 126

6.3.1 Carbon Neutral Fruits: The Dole Carbon Neutral Strategy for Bananas and Pineapples 127
6.3.2 Carbon Neutral Coffee: The Coopedota Case Study 129
6.3.3 Carbon Neutral Meat: The Five Founders Case Study 131
References 136

7 Carbon Neutrality as Business: Decoupling Growth from Emissions While Managing Climate Risks 139
7.1 The Role of consumers' Food Choices in Mitigating Carbon Emissions 139
7.2 Consumers' Willingness to Pay for Climate Friendly and Climate Neutral Products 141
7.3 Market drivers and Businesses Opportunity for Agri-Food Companies to Move Towards Lower Carbon Emissions 143
7.4 Barrier and Risk for Suppliers to Join the Carbon Neutral Chain 145
7.5 Different Development Models for Carbon Neutrality 148
7.5.1 Government-Driven: The French Label Bas Carbone 148
7.5.2 Sectoral Legislation-Driven: The ICAO Model 150
References 152

8 Towards a Carbon Labelled World? 155
8.1 Consumers Perceptions and Use of Carbon Labelling 155
8.2 Carbon labels in Relation to Others Greening Themes and Labels 158
8.3 Carbon Labels: Does Design Matter? 161
8.4 How to Communicate Climate change and Products Carbon Information 165
8.5 The Role of retailers and Green Public Procurements (GPP) in Boosting Carbon Reduction and Carbon Neutral Labels 170
References 172

Conclusions and Recommendations 177
References 184

Index 185

List of Figures

Fig. 1.1 European citizens main environmental concerns
Source: Authors' elaboration on European Commission (2017) data 2
Fig. 1.2 Perceptions of climate change as a global problem
Source: Authors' elaboration on European Commission-Directorate General (2019) data 3
Fig. 1.3 Personal action to tackle climate change
Source: Authors' elaboration on European Commission-Directorate General (2019) data 3
Fig. 1.4 Types of individual action
Source: Authors' elaboration on European Commission-Directorate General (2019) data 4
Fig. 1.5 A climate-neutral Europe by 2050
Source: Authors' elaboration on European Commission-Directorate General (2019) data 5
Fig. 1.6 Climate change international Agreements
Source: Authors' elaboration on Barilla Center for Food and Nutrition (2016) 6
Fig. 1.7 IPCC special Report 7
Fig. 1.8 European Strategy for Carbon Neutrality 8
Fig. 1.9 Carbon neutrality Milestones
Source: Authors' own elaboration 9
Fig. 1.10 INDCs and NDCs
Source: https://www.activesustainability.com/climate-change/what-are-ndcs-why-are-important-climate-change/ 12
Fig. 1.11 Main differences between NDCs and LTSs
Source: Kelly and Franse (2019) 13
Fig. 1.12 Types of mitigation related MRV
Source: Singh et al. (2016) 14

Fig. 1.13 Article 13 of the Paris Agreement: transparency of action and support
Source: https://unfccc.int/process-and-meetings/transparency-and-reporting/the-big-picture/what-is-transparency-and-reporting 15
Fig. 1.14 Transition from the MRV framework to the Enhanced Transparency Framework and accounting
Source: Wartmann et al. (2018) 16
Fig. 1.15 Countries' Carbon Neutral policies
Source: Energy and Climate Intelligence Unit (2019) 17
Fig. 1.16 Countries with a net zero commitment by 2050
Source: Energy and Climate Intelligence Unit (2019) 18
Fig. 1.17 Boundary of corporate climate neutrality targets
Source: Pineda and Faria (2019) 21
Fig. 1.18 Carbon neutral definitions 23
Fig. 1.19 Main step to reach carbon neutrality 24
Fig. 1.20 Main step to reach carbon neutrality 24

Fig. 2.1 Agriculture and Forestry and land use GHG emissions
Source: Richards et al. (2019) 30
Fig. 2.2 GHG emissions from food production
Source: Poore and Nemecek (2018) 31
Fig. 2.3 GHG emissions from food
Source: Poore and Nemecek (2018) 32
Fig. 2.4 GHG emissions from different foods
Source: Poore and Nemecek (2018) 33
Fig. 2.5 Paris Agreement and Nationally Determined Contributions in the Agri-food sector
Source: Wieben (2019) 34
Fig. 2.6 Paris Agreement and NDCs in the Agri-food sector - Mitigation and adaptation measures
Source: Wieben (2019) 34
Fig. 2.7 A snapshot of mitigation policy progress in the agriculture sector
Source: OECD (2019) 35
Fig. 2.8 Mitigation potential of response options in 2020–2050, measured in GtCO2-eq $yr.^{-1}$
Source: Shukla et al. (2019) 38

Fig. 2.9 Potential global contribution of response options to mitigation, adaptation, combating desertification and land degradation, and enhancing food security**Panel A** shows response options that can be implemented without or with limited competition for land, including some that have the potential to reduce the demand for land. Co-benefits and adverse side effects are shown quantitatively based on the high end of the range of potentials assessed. Magnitudes of contributions are categorized using thresholds for positive or negative impacts. Letters within the cells indicate confidence in the magnitude of the impact relative to the thresholds used (see legend). Confidence in the direction of change is generally higher
Source: Shukla et al. (2019) 39
Fig. 2.10 Economic mitigation potential for the AFOLU sectorMitigation potential for the AFOLU sector, plotted using data from AR4 (Nabuurs et al., 2007; Smith et al., 2007). Whiskers show the range of estimates (+/− 1 standard deviation) for agricultural options for which estimates are available
Source: Smith et al. (2014) 40
Fig. 2.11 Economic mitigation potential by sector in 2030 estimated from bottom-up studies
Source: IPCC (2007) 41
Fig. 2.12 Drivers of GHG emissions growth and mitigation, 2015–2050, 2 °C scenario, World
Source: Keramidas et al. (2018) 42
Fig. 2.13 UN-REDD Programme
Source: UN-REDD Programme (2015) 43
Fig. 2.14 The 4 per 1000 initiative
Source: French Ministry for Agriculture & Food (2015) 44
Fig. 2.15 The plant for the planet and the Bonn challenge initiatives
Source: https://www.plant-for-the-planet.org; https://www.bonnchallenge.org 45
Fig. 2.16 Fairtrade Carbon Credits
Source: https://www.fairtrade.net/product/carbon-credits 46
Fig. 2.17 Rainforest Alliance Logo
Source: https://www.rainforest-alliance.org/articles/understanding-tools-to-fight-climate-change 47

Fig. 3.1 Life Cycle Assessment and Footprint
Source: Barilla Center for Food and Nutrition (2016) 52
Fig. 3.2 Different GHG protocol standards
Source: http://ghgprotocol.org 57
Fig. 3.3 The GHG protocol for guidance on carbon removals and land use
Source: http://ghgprotocol.org 58

Fig. 3.4 Scope of the Organizational Carbon footprint according to the GHG protocol
Source: World Resources Institute and the World Business Council for Sustainable Development (2011) in the GHG protocol 59

Fig. 4.1 Types of greenwash
Source: Horiuchi et al. (2009) 64
Fig. 4.2 McKinsey survey' results -Part 1
Source: Bernow et al. (2019) 67
Fig. 4.3 McKinsey survey' results – Part 2
Source: Bernow et al. (2019) 68

Fig. 5.1 Carbon Neutrality Guiding Scheme
Source: Authors' own elaboration 72
Fig. 5.2 Tesco and Casino logos 73
Fig. 5.3 Walmart and Heineken logos 73
Fig. 5.4 Alter Eco logo 74
Fig. 5.5 Max Burger and Unilever logos 74
Fig. 5.6 Logos of companies that have declared a Carbon Neutral Commitment 75
Fig. 5.7 Logos of companies that have issued a Carbon Neutrality Statement 75
Fig. 5.8 Logos of companies that have adopted a Carbon Neutral Label. 75
Fig. 5.9 The Gold Standard logo 77
Fig. 5.10 The Verra logo 77
Fig. 5.11 The UNFCCC logo 78
Fig. 5.12 The Plan Vivo logo 79
Fig. 5.13 The Social Carbon logo 79
Fig. 5.14 The CCBS logo 79
Fig. 5.15 Transacted Voluntary Carbon offset volume, value and weighted average price by project category, 2017 and 2018
Source: Forest Trends' Ecosystem Marketplace (2019) 82
Fig. 5.16 Forest Trends' Ecosystem Marketplace Carbon offset project types and categories, pre-2018
Source: Forest Trends' Ecosystem Marketplace (2019) 83
Fig. 5.17 Forest Trends' Ecosystem Marketplace Carbon offset project types and categories, in 2018
Source: Forest Trends' Ecosystem Marketplace (2019) 84
Fig. 5.18 Offsetting and Insetting
Source: Tipper et al. (2009) 85
Fig. 5.19 International Insetting Platform Logo and members
Source: International Insetting Platform (2017) 86
Fig. 5.20 Nespresso logo 87
Fig. 5.21 Ben & Jerry's logo 87

Fig. 5.22 Chocolats Halba logo ... 88
Fig. 5.23 Insetting benefits
Source: https://www.purprojet.com/presentation-what-is-insetting/ ... 89
Fig. 5.24 Carbon neutral commitment of companies and brands
Source: Authors' own elaboration ... 93
Fig. 5.25 Comparison of standards and technical regulations
Source: Rousset et al. (2015) ... 97
Fig. 5.26 Core characteristics of international standards, regulations, and private standards
Source: International Centre for Trade and Sustainable Development (ICTSD) (2017) ... 98
Fig. 5.27 Voluntary standards for Carbon Neutrality process
Source: Authors' own elaboration ... 98
Fig. 5.28 PAS 2060: 2014 Specification for the demonstration of carbon neutrality
Source: BSI (2014) ... 99
Fig. 5.30 The four steps towards CN
Source: Authors' own elaboration ... 100
Fig. 5.29 Carbon Neutrality Equation
Source: Carbon Clear (2011) ... 100
Fig. 5.31 Key steps for reaching and maintaining the carbon neutrality according to the reference standard
Source: Authors' own elaboration ... 101
Fig. 5.32 Climate Active Carbon Neutral Standard for Organisations
Source: https://www.climateactive.org.au/be-climate-active ... 102
Fig. 5.33 Climate Active Carbon Neutral labels
Source: https://www.climateactive.org.au/be-climate-active ... 103
Fig. 5.34 Standards systems
Source: Loconto and Dankers (2014) ... 104
Fig. 5.35 Overview of standards and certification systems
Source: Authors' own elaboration ... 106
Fig. 5.36 Gran Cereale Standards and verification system
Source: Authors' own elaboration ... 107

Fig. 6.1 Agri-food supply chain actors
Source: KPMG (2013) ... 116
Fig. 6.2 Shares of food supply chain stages in food-related GHG emissions (excluded land use change)
Source: Noleppa (2012) ... 116
Fig. 6.3 CSR practices and agri-food supply chain
Source: Murphy-bokern and Kleemann (2014) ... 117
Fig. 6.4 BCFN Double Pyramids
Source: Barilla Center for Food and Nutrition (2016) ... 122

Fig. 6.5 Sources used to build the pyramids
Source: Barilla Center for Food and Nutrition (2016) 123
Fig. 6.6 Number of sources used to build the pyramids database (by years and for different types of footprint)
Source: Barilla Center for Food and Nutrition (2016) 124
Fig. 6.7 Distribution of data by source type and distribution of bibliographical sources relative to environmental impacts
Source: Barilla Center for Food and Nutrition (2016) 124
Fig. 6.8 The Water Footprint and Environmental Footprint pyramidsSourc: Barilla Center for Food and Nutrition (2016) 125
Fig. 6.9 The Carbon Footprint Pyramid
Source: Barilla Center for Food and Nutrition (2016) 125
Fig. 6.10 GHG emissions and the food chain
Source: Barilla Center for Food and Nutrition (2016) 126
Fig. 6.11 Dole logo
Source: www.dolecrs.com 127
Fig. 6.12 Dole's bananas Carbon footprint
Source: Soil & More International B.V. (2010) 128
Fig. 6.13 Coopedota carbon neutral initiative
Source: www.coopedota.com 129
Fig. 6.14 Coopedota steps towards carbon neutrality
Source: Authors own elaboration based on Jiménez et al. (2013) 130
Fig. 6.15 Five Founders Carbon Neutral logo
Source: https://fivefounders.com.au/our-difference/#environment 132
Fig. 6.16 Five Founders Carbon Neutrality strategy
Source: https://fivefounders.com.au/our-difference/#environment 132
Fig. 6.17 Five Founders LCA Boundary
Source: NAPCO' Australian Government Carbon Neutral Program Public Disclosure Summary (2019) 134
Fig. 6.18 Emissions of Five Founders (2018)
Source: NAPCO' Australian Government Carbon Neutral Program Public Disclosure Summary (2019) 135
Fig. 6.19 NAPCO offset strategy
Source: NAPCO' Australian Government Carbon Neutral Program Public Disclosure Summary (2019) 136

Fig. 7.1 Pinkse and Busch (2013) framework
Source: Pinkse and Busch (2013) 143
Fig. 7.2 Reasons for corporate climate actions
Source: We Mean Business (2015) 146
Fig. 7.3 Label Bas Carbone Logo and Development Model 149
Fig. 7.4 Carbon Agri and CAP2'ER Logos 150

Fig. 7.5 The CORSIA timeline
Source: Forest Trends' Ecosystem Marketplace (2019) 151
Fig. 8.1 Main Food Labels
Source: Authors' own elaboration 159
Fig. 8.2 The green-yellow-red combination
Source: Sharp and Wheeler (2013) 162
Fig. 8.3 The green-yellow-black combination
Source: Sharp and Wheeler (2013) 162
Fig. 8.4 Carbon trust label 163
Fig. 8.5 Carbon scale
Source: Hartikainen et al. (2014) 163
Fig. 8.6 Carbon scale and GHG emissions
Source: Feucht and Zander (2018), Meyerding et al. (2019) 163
Fig. 8.7 Label on the right visual field with horizontal allocation
Source: Zhou et al. (2019) 164
Fig. 8.8 Carbon trust, carbon label and carbon reduction label
Source: Meyerding et al. (2019) 165
Fig. 8.9 Socio-demographic characteristics of the sample
Source: Klein and Menrad (2016) 166
Fig. 8.10 Preferred information and supporting instruments
Source: Klein and Menrad (2016) 167
Fig. 8.11 Six barriers on food eco-labels effectiveness
Source: Grunert (2011) 168

List of Tables

Table 1.1 Countries that have submitted long-term low greenhouse gas emission development strategies 10
Table 1.2 Existing MRV arrangements for developed countries and developing countries 15
Table 1.3 Carbon neutrality definitions 20

Table 3.1 Comparison of Carbon related standards 55

Table 5.1 Comparisons of the main Carbon offset standards 80
Table 5.2 Voluntary offset market trends 81
Table 5.3 Gran Cereale Case study 106
Table 5.4 Carbon labels 109

Table 6.1 Measures to reduce GHG emission along the agri-food supply chain 118
Table 6.2 Coopedota amount of certified coffee, the carbon footprint and emission to be offset 130
Table 6.3 Internal and external certification cost 131

Table 7.1 Synthesis of the results from studies investigating WTP for carbon labeled products 141
Table 7.2 Risks of carbon labelling for suppliers in developing countries 148

Table 8.1 The Italian GPP Action Plan 171

Chapter 1
Carbon Neutrality: If Not Now, When?

The first chapter offers an overview of the climate change and carbon neutrality policies of the various countries. A particular focus is dedicated to agri-food policies related to climate change and different mitigation projects developed in the sector.

1.1 Citizens' Perception of Climate Change Seriousness

In the last few years climate change has become one of the main global issues (Braun 2010) and one of the most challenging communication issues for marketers in the commercial and public sector (Peattie et al. 2009). In recent times, the intensification of extreme climatic phenomena, the fires in the Amazon Forest, the problem of plastic, the youth movement of Greta Thunberg have raised an unprecedented media attention on the phenomenon of Climate Change. This has led many citizens, especially young people, to urge governments to find effective and rapid solutions to global warming. In this sense, the academic literature is struggling to keep up in photographing and analyzing changes in consumer awareness and concern about environmental problems. In fact, there are still few academic studies that analyze the impact of these phenomena on the public opinion and on the predisposition of consumers to change their habits to decrease their impact on the environment. Probably the next studies that will be published will examine in depth the implications that these phenomena have had in the environmental awareness of citizens, especially as far as millennials and generation Z are concerned. All this being considered, following the current news, one cannot help but notice that citizens are clamoring for substantial changes in the way our economy produces and consumes goods and services. Moreover, the Greta Thunberg international youth movement against climate change "#FridaysforFuture" has stirred the consciences of citizens and has encouraged global awareness of the negative consequences of climate change. Several scholars have found that consumers are increasingly concerned about their impacts on the environment and on the climate (Sirieix et al. 2013;

A. Acampora et al., *Carbon Neutrality in the Agri-food Sector*,
https://doi.org/10.1007/978-3-030-88048-4_1

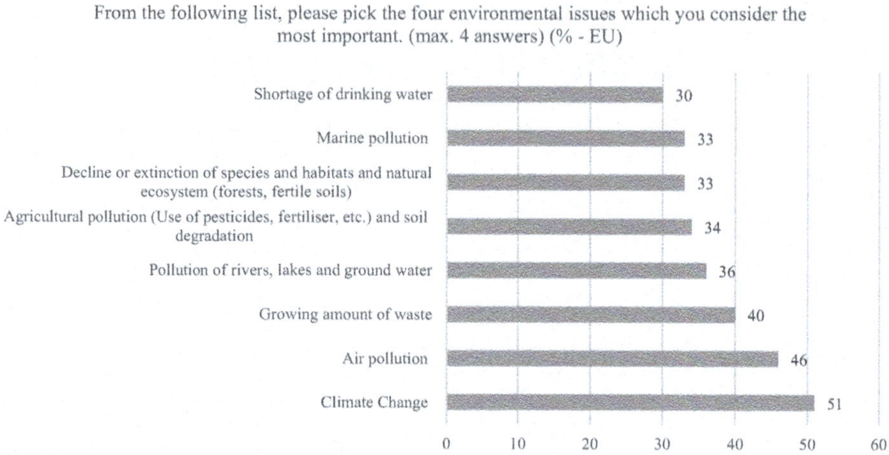

Fig. 1.1 European citizens main environmental concerns
Source: Authors' elaboration on European Commission (2017) data

Thøgersen and Nielsen 2016; Vanclay et al. 2011). The European Commission (2017) elaborated a report on the attitudes of European citizens towards the environment. According to the results of the respondents (27,881 EU citizens) consider climate change (51%), air pollution (46%) and the growing amount of waste (40%) the most important environmental issues (Fig. 1.1).

Particularly, a recent report by European Commission- Directorate General (2019) reported results from a survey of the Eurobarometer carried out in the 28 Member States of the European Union in April 2019 with 27,655 respondents from different social and demographic groups. From the report emerged that almost one quarter (23%) of the respondents mention climate change as the most serious problem facing the world, positing itself as second most serious problem after the poverty, hunger and lack of drinking water (27%). The most significant change since 2017 appeared, indeed, in the climate change seriousness perception, with an increase of 11 percentage points (Fig. 1.2).

Moreover, 60% of respondents stated that they have taken any personal action to fight climate change in the past 6 months. The proportion who says they have taken personal action in the past 6 months has increased considerably since 2017 (+11 percentage points) and is now at its highest ever point (Fig. 1.3).

Regarding the type of actions undertaken by the respondents, 75% stated that they have tried to reduce their waste and regularly separate it for recycling, while 62% said they try to cut down on their consumption of disposable items whenever possible. 48% said that when buying a new household appliance, lower energy consumption is an important factor in their choice. Particularly, almost one in five (18%) consider the carbon footprint of their food purchases and sometimes adapt their shopping accordingly, while 16% say they have installed equipment in their home to control and reduce their energy consumption. Generally, 93% of respondents have taken at least one personal action that helps tackle climate change. In fact,

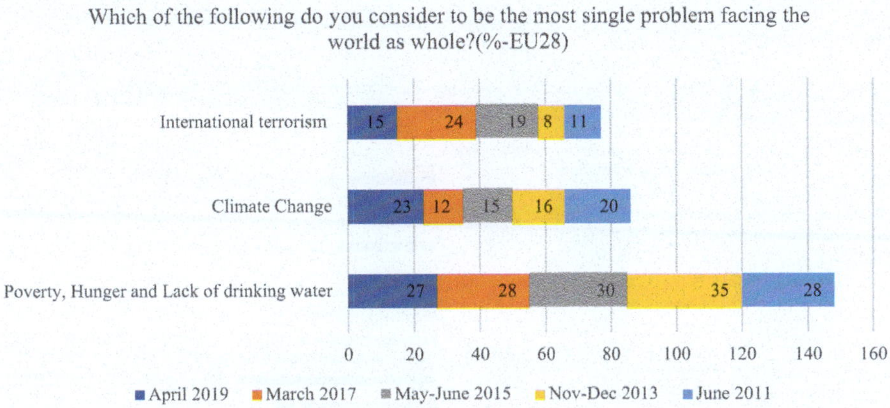

Fig. 1.2 Perceptions of climate change as a global problem
Source: Authors' elaboration on European Commission- Directorate General (2019) data

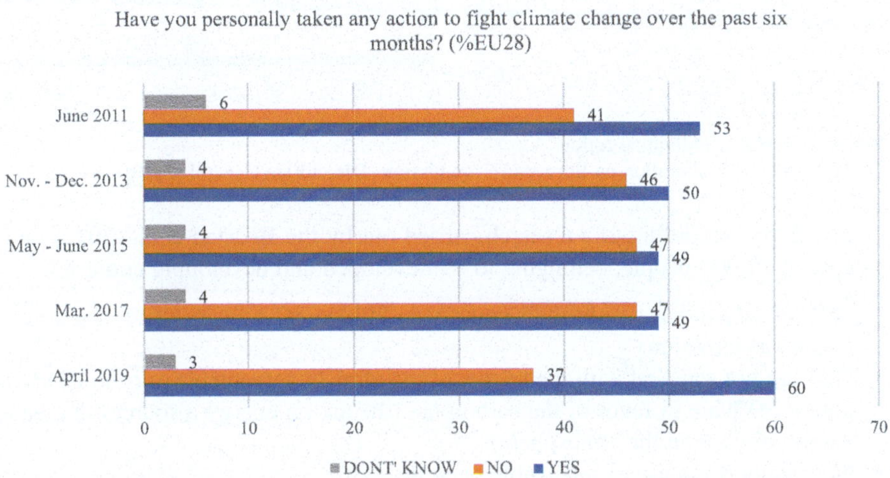

Fig. 1.3 Personal action to tackle climate change
Source: Authors' elaboration on European Commission- Directorate General (2019) data

although 60% of respondents say they have personally taken action to fight climate change over the past 6 months (Fig. 1.3), when asked about specific actions almost all respondents (93%) have taken at least one personal action that helps fight climate change, 6% say that they have not taken any of these actions, while 3% have taken some other action (Fig. 1.4).

Furthermore, more than nine in ten respondents (92%) agree that greenhouse gas emissions should be reduced to a minimum while offsetting the remaining emissions, in order to make the EU economy climate neutral by 2050. In fact, the majority (54%) "totally agree". Only 5% disagree, while 4% say they don't know (European Commission- Directorate General 2019) (Fig. 1.5).

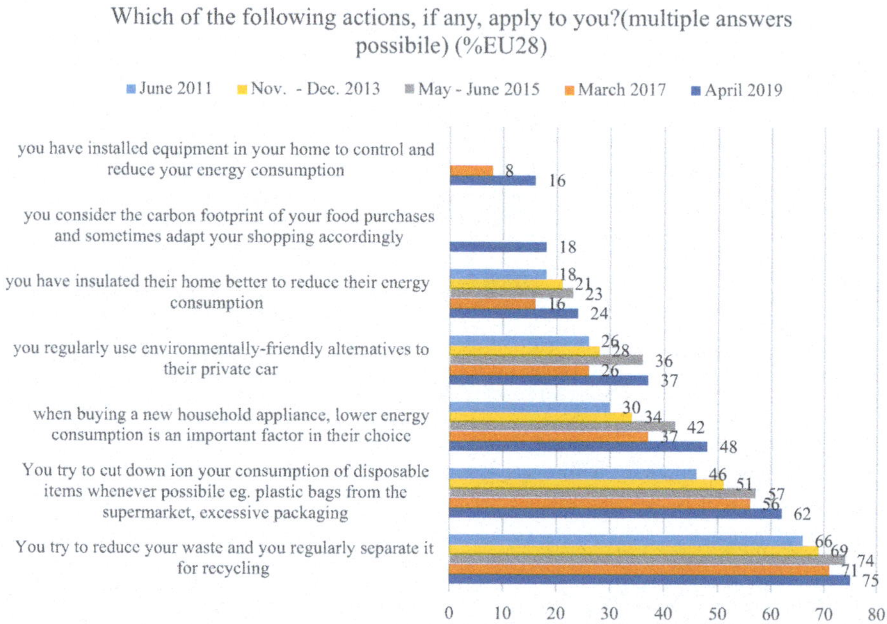

Fig. 1.4 Types of individual action
Source: Authors' elaboration on European Commission- Directorate General (2019) data

Moreover, according to a research carried out by the Bedsted et al. (2015) on a sample of 10,000 people, belonging to 79 developed and developing countries:

- 78% of respondents want their country to implement policies to reduce greenhouse gas emissions
- 70% are not satisfied with the results of the climate agreements prior to COP21.
- almost 90% are in favor of the carbon tax (the tax on energy resources that emit carbon dioxide in the atmosphere),
- 56% support the use of renewable energies
- 46% would support the development of environmentally friendly technologies
- 97% of want a Paris agreement to include a long-term goal for zero emissions at the end of this century. Two-thirds of participants want this goal to be legally binding for all countries. 17% think it should be legally binding only for developed and emerging nations.

1.2 Climate Change and Carbon Neutrality Policies

Contrasting climate change would require significant and constant reductions in greenhouse gas (GHG) emissions (Intergovernmental Panel on Climate Change 2014). At the United Nations Conference on Environment and Development held

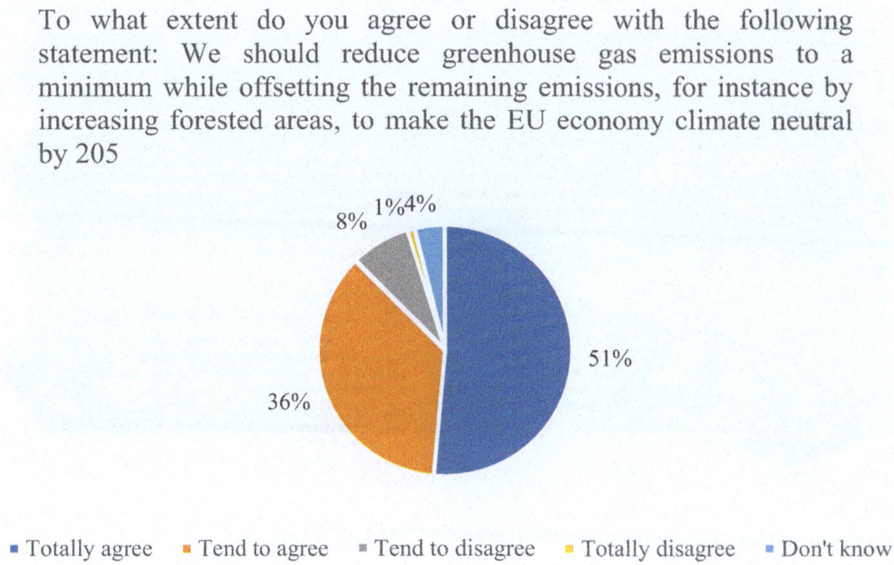

Fig. 1.5 A climate-neutral Europe by 2050
Source: Authors' elaboration on European Commission- Directorate General (2019) data

in Rio de Janeiro in 1992, the United Nations Framework Convention on Climate Change (UNFCCC) was adopted with the aim of stabilizing the concentrations of greenhouse gases in the atmosphere at sustainable level and countering such serious environmental consequences. Following ratification by the 50th State, it entered into force in 1994. Signatory States meet at regular intervals during the so-called COP (Conferences of the Parties) to discuss on further actions on climate protection. In 1997, this meeting was held in Kyoto, Japan, in which the "Kyoto Protocol", the first document with legally binding obligations for CO2 limits and reductions, was adopted by ratified countries. The period of applicability was set for the years 2008 to 2012 (first obligation period) and from 2013 to 2020 (second obligation period). Under the Kyoto protocol, industrialized nations have committed to reducing their annual hydrocarbon emissions, measured in six types of greenhouse gases of varying quantities, for an average reduction of 5.2% by 2012 compared to the 1990. This amount of emission reductions would amount to a 29% reduction in greenhouse gas emissions that would otherwise have occurred. However, each country had its goal. For example, when the agreement was concluded, the members of the European Union (EU) had the objective of reducing emissions by 8% by 2012; the United States had a 7% reduction target and Canada's target was 6%. The last international agreement on climate change has been negotiated in Paris in December 2015 at the United Nations Climate Change Conference in which the so called "Paris Agreement", on the reduction of the GHG emissions, was signed by several countries. In the agreement, the signing parties committed themselves to limiting the increase of global warming to less than 2 degrees Celsius compared to pre-industrial

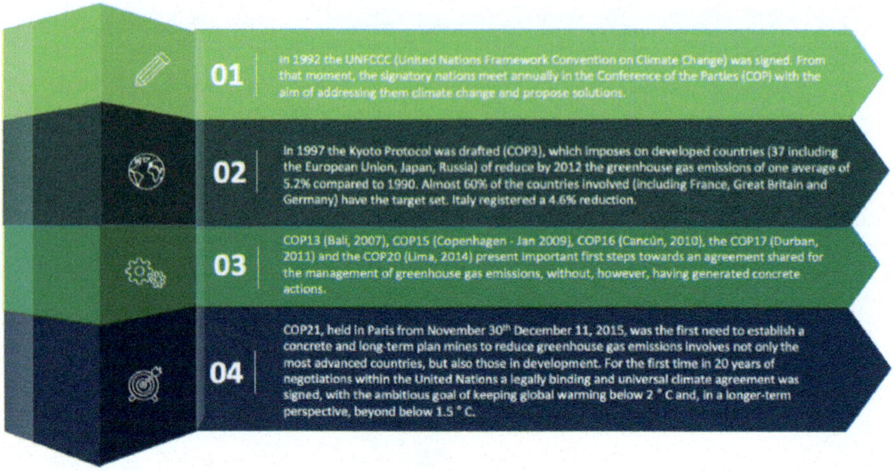

Fig. 1.6 Climate change international Agreements
Source: Authors' elaboration on Barilla Center for Food and Nutrition (2016)

levels (United Nations Framework Convention on Climate Change 2015). To meet these climate change targets a widespread public commitment, coming from both governments, industry, and citizens, is necessary (Feucht and Zander 2018). Figure 1.6 summarizes the main climate change international agreements.

In 2014 the UNEP issued a communication in which they stated that in order to limit global temperature rise to 2 °C and head off the worst impacts of climate change, global carbon neutrality should be attained by mid-to-late century (United Nations Environment Programme (UNEP) 2014). In the same year the UN climate summit was held in New York. The Summit brought together 100 Heads of State, together with government ministers and leaders from international organizations, business, finance, civil society and local communities, to mobilize the political support and momentum necessary to reach a global agreement on climate change in 2015 and galvanize action on the ground across all sectors. In that occasion the United Nations has made a firm commitment that we will make all the United Nations systems carbon neutral by 2020 (United Nations 2014). In 2015, the B Team Leaders, a non-profit organization co-founded Sir Richard Branson and Jochen Zeitz, went to COP21 in Paris to show business support for a long-term goal embedded in the Paris Agreement and to urge governments to reach an agreement towards net-zero greenhouse gas emissions by 2050. In this context the Carbon neutrality was born. The concept has received considerable interest in recent years, as a tool for the decarbonization of the world economy, by governments, businesses and consumers. This concept, even if not with the well-know "carbon neutrality"expression, it was already in the article 2 of the 1992 United Nations Framework Convention on Climate Change (UNFCCC) as a directive to signatories parties to achieve "*stabilization of greenhouse gas concentrations in the atmosphere at a level that would prevent dangerous anthropogenic interference with the climate*

Fig. 1.7 IPCC special Report

system"(UNFCCC 1992). However, it with the article 4.1 of the more recent Paris Agreement, adopted in 2015, that carbon neutrality has been introduced into international agreements (Vallejo et al. 2018). The article 4, paragraph 1, refers to need to: "*achieve a balance between anthropogenic emissions by sources and removals by sinks of greenhouse gases in the second half of this century, on the basis of equity, and in the context of sustainable development and efforts to eradicate poverty*" (United Nations Framework Convention on Climate Change 2015).

Moreover, as part of the decision of the 21st Conference of Parties of the UNCCC to adopt the Paris Agreement, the Intergovernmental Panel on Climate Change (IPCC) was called to produce, in 2018, a Special Report on global warming of 1.5 °C above pre-industrial levels and related global greenhouse gas emission pathways to ensure scientific consistency to governmental actions. The IPCC report "Global warming of 1.5 °C", released in 2018, concluded that global emissions need to reach net zero around mid-century to give a reasonable chance of limiting warming to 1.5 °C, and if this target will be meet in 2070 we should be able to contain global warming below 2 °C (IPCC 2018a; Masson-Delmotte et al. 2018) (Fig. 1.7).

Following the Paris Agreement and the IPCC report, also the European Union, in November of 2018, released its strategic long-term vision for a climate neutral economy by 2050 with the communication "*A Clean Planet for all – A European strategic long-term vision for a prosperous, modern, competitive and climate neutral economy*" (28/11/2018 – COM (2018) 773) (European Commission 2018) (Fig. 1.8).

With this communication the European Union presented its vision for achieving net-zero greenhouse gas emissions by 2050 through a socially fair transition in a cost-efficient manner.

The European vision for a climate neutral Europe is mainly based on 7 strategic building blocks (Directorate-General for Climate Action -European Commission 2019):

- maximize the benefits of energy efficiency, including zero emission buildings;
- maximize the deployment of renewables and the use of electricity to fully decarbonize Europe's energy supply;
- embrace clean, safe and connected mobility;

Fig. 1.8 European Strategy for Carbon Neutrality

- a competitive EU industry and the circular economy as a key enabler to reduce GHG emissions;
- develop an adequate smart network infrastructure and interconnections;
- reap the full benefits of bio economy and create essential carbon sinks;
- tackle remaining CO_2 emissions with Carbon Capture and Storage (CCS).

Figure 1.9 summarizes the main milestones towards carbon neutrality.

1.2.1 Long-Term Low Greenhouse Gas Emission Development Strategies and Nationally Determined Contributions (NDCs)

In paragraph 19 of the Article 4 of the Paris Agreement, it is prescribed that all the signatory parties should formulate and communicate by 2020 their long-term low greenhouse gas emission development strategies.

Long-term low greenhouse gas emission development strategies (LT-LEDS) are national, subnational or supranational long-term strategies for envisioning low-emission development while identifying national priorities that are in line with decarbonization pathways. LT-LEDS are voluntary strategies that guide governments as they consider climate change mitigation and national socio-economic development in an integrated and strategic manner (Rocha and Falduto 2019). At the moment the countries that have submitted the long-term strategies for post Paris are regularly updated on the UNFCC website (UNFCC 2019a) (Table 1.1).

In the lead up to the historic Paris Agreement on climate change, adopted in 2015, more than 160 countries and the European Union submitted their own plans to address climate change, known as Intended Nationally Determined Contributions

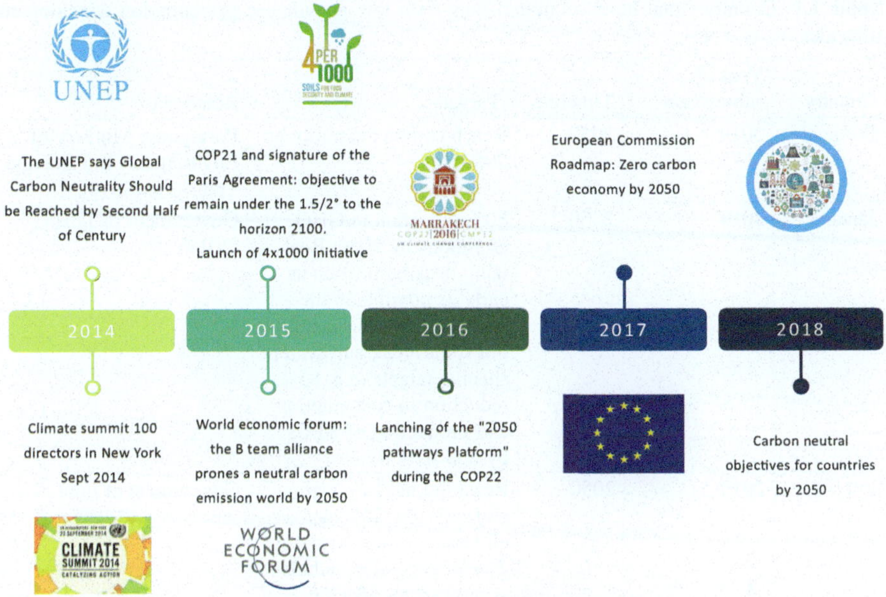

Source: Authors' own elaboration.

Fig. 1.9 Carbon neutrality Milestones
Source: Authors' own elaboration

(INDCs). According to the global climate pact, a country's INDC is converted to a Nationally Determined Contribution (NDC) when it formally joins the Paris Agreement by submitting an instrument of ratification, acceptance, approval or accession, unless a country decides otherwise. **Nationally Determined Contributions** (NDCs) are **countries' commitments** to international climate action, demonstrating how countries are willing to contribute to the collective global goals of the Paris Agreement. For the most part, the first round of NDCs targeted 2025 or 2030.

Most NDCs contain economy-wide quantified emissions reduction targets (for example, reduce 25% below 1990 levels by 2030), which send a signal about where a country's emissions are headed. Many NDCs also contain non-GHG targets, such as a certain percentage of renewable energy generation or amount of forest restoration to be achieved. Most NDCs also articulate specific actions or policies they will pursue to achieve their targets (Kelly and Franse 2019). The Paris Agreement established a series of 5-year cycles to increase ambition, including through NDCs that would grow more ambitious over time. Countries also specifically agreed on 2020 as a critical next step in the Paris process. Countries with an NDC time frame ending up to 2025 are requested to communicate new ones by 2020 (United Nations Framework Convention on Climate Change 2015).

Those countries whose NDCs have a time frame ending up to 2030, in turn, are requested to communicate or update their NDCs by 2020. This guidance refers to the process of developing new or updated NDCs as "NDC enhancement "(WRI 2019) (Figs. 1.10 and 1.11).

Table 1.1 Countries that have submitted long-term low greenhouse gas emission development strategies

Country	Year of submission	Timeline	Targets	Reference
Portugal	2019	2050	Reach carbon neutrality by 2050	Portuguese Ministry of Environment and Energy Transition (2019)
Japan	2019	2050	80% reduction in GHG emissions by 2050; Realizing decarbonization as early as possible in the second half of this century; For some areas, sub-sector-specific targets (e.g. 85% reduction in consumption and production of HFCs by 2036).	Government of Japan (2019)
Fiji	2019	2050	Reaching net zero carbon emissions by 2050 across all sectors of the economy; The strategy also includes several sector-specific quantitative targets	Government of Fiji (2018)
Republic of the Marshall Islands	2018	2050	Reaching net zero carbon emissions by 2050; Reaching 100% renewable energy; Facilitating adaptation and climate change resilience.	The Republic of the Marshall Islands (2018)
Ukraine	2018	2050	Decarbonization of the energy system; Increase in the volumes of carbon absorption; Reduction in GHG emissions such as methane gas and nitrogen oxide.	Ukraine Ecology and Natural Resources Ministry (2018)
United Kingdom	2018	2050	Sets out plans and proposals to meet the 'carbon budgets' set to date – which require a 57% reduction in GHG emissions, relative to 1990 levels over the period 2028–32. Sets out scenarios for meeting the 2050 target (an 80% reduction on 1990 levels – at the time of writing). Sets out indicative sectoral pathways to 2032 alongside a range of	UK Department for Business Energy and Industical Strategy; HM Government (2017)

(continued)

Table 1.1 (continued)

Country	Year of submission	Timeline	Targets	Reference
			sector-specific goals, ambitions and policies.	
Czech Republic	2018	2050	Reduce national emissions by 2020 by at least 32 Mt CO2-eq in comparison with 2005; Reduce national emissions by 2030 by at least 44 Mt CO2-eq in comparison with 2005; Pursue the indicative level of 70 Mt CO2-eq of emissions in 2040; Pursue the indicative level of 39 Mt CO2-eq of emissions in 2050.	Ministry of the Environment of the Czech Republic (2017)
France	First submission 2016; resubmission 2017	2050	75% reductions in GHG emissions. In 2050 compared to 1990 5-year carbon budgets until 2028	Ministère de la transition écologique et solidaire (2015)
Benin	2016	2025	Reinforce resilience of local communities;12Mt GHG reduction and 163Mt GHG sequestered by 2025; Reinforce protection of vulnerable communities.	Benin Ministère du Cadre de vie et du Développement Durable (2016)
United States	2016	2050	Achieving at least 80% GHG emissions reductions by 2050	The White House (2016)
Mexico	2016	2050	Reaching at least 50% of energy generation from clean sources in 40 years; Achieving 50% emissions reduction compared to 2000 in 40 years; Several other milestones and targets	Government of Mexico (2016)
Germany	First submission 2016; resubmission 2017	2050	80–95% reduction in GHG emissions compared to 1990	Federal Ministry for the Environment Nature Conservation Building and Nuclear Safety (BMUB) (2016)
Canada	2016	2050	80% reduction in GHG emissions compared to 2005	Government of Canada -Environment and Climate Change Canada (2016)

Source: UNFCC (2019a)

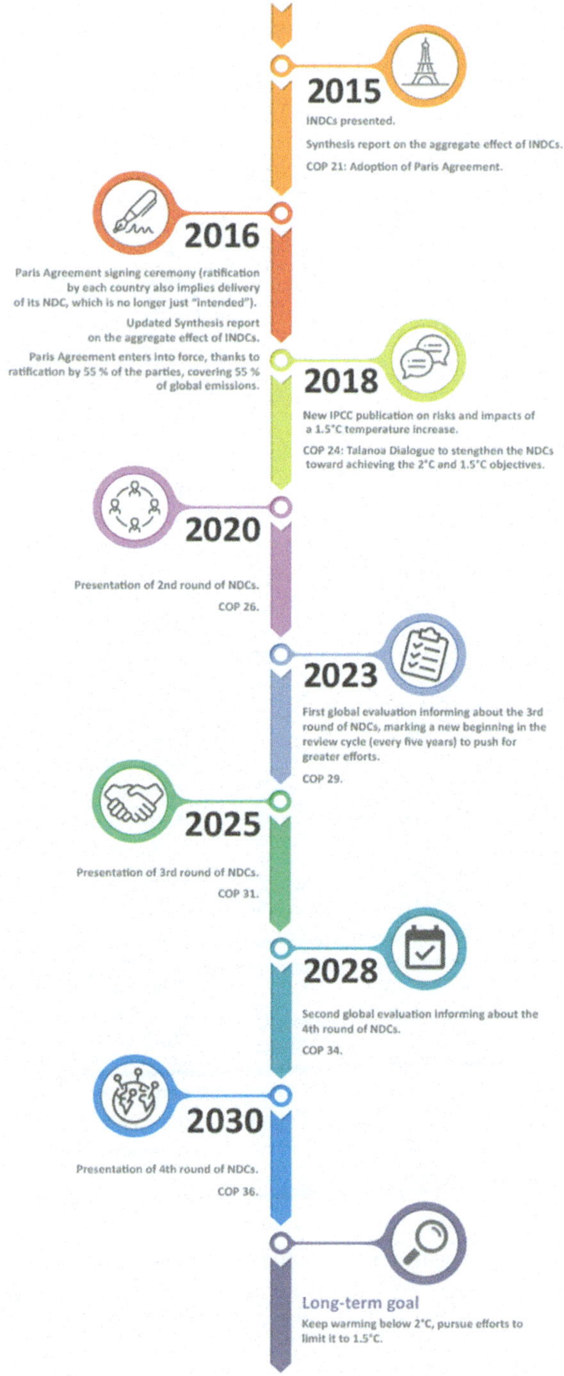

Fig. 1.10 INDCs and NDCs
Source: https://www.activesustainability.com/climate-change/what-are-ndcs-why-are-important-climate-change/

	NDCs	LTSs
Planning horizon	Approximately 10-15 years	Approximately 30 years
Frequency	Every 5 years	Not specified
Targets	"Developed country Parties should continue taking the lead by undertaking economy-wide absolute emission reduction targets. Developing country Parties should continue enhancing their mitigation efforts, and are encouraged to move over time towards economy-wide emission reduction or limitation targets in the light of different national circumstances."	Not specified. In practice, all of the countries whose LTS extent to 2050 have included one or several quantitative GHG reduction targets.
Accounting	"Parties shall account for their nationally determined contributions. In accounting for anthropogenic emissions and removals corresponding to their nationally determined contributions, Parties shall promote environmental integrity, transparency, accuracy, completeness, comparability and consistency, and ensure the avoidance of double counting, in accordance with guidance adopted by the [CMA]."	Not specified
Reporting	"In communicating their nationally determined contributions, all Parties shall provide the information necessary for clarity, transparency and understanding in accordance with decision 1/CP.21 and any relevant decisions of the Conference of the Parties serving as the meeting of the Parties to this Agreement." "Each Party shall regularly provide...Information necessary to track progress made in implementing and achieving its nationally determined contribution under Article 4."	Not specified
Legal character	"Each Party shall prepare, communicate and maintain successive nationally determined contributions that it intends to achieve"	"All Parties should strive to formulate and communicate long-term low greenhouse gas emission development strategies"

Source: (Kelly and Franse, 2019).

Fig. 1.11 Main differences between NDCs and LTSs
Source: Kelly and Franse (2019)

1.2.2 The UNFCCC's Measurement, Reporting and Verification (MRV) Framework and Enhanced Transparency Framework (ETF)

The term measurement, reporting and verification (MRV) framework has been developed gradually over the period 2004–2013 in different sessions COP, but appeared for the first time at COP 13 hold in Bali in the context of climate change mitigation policy, were the Bali Action Plan (2007) was developed, which called for "*measurable, reportable, and verifiable nationally appropriate mitigation commitments or actions*" and stated that they should be "*supported and enabled by technology, financing and capacity-building, in a measurable, reportable and verifiable manner*" (UNFCCC 2008). Subsequently, efforts have been made to fill in the details and define what should be measured, reported, and verified, how, by whom, and for what purpose (Singh et al. 2016). The MRV framework includes the following steps and procedures (Singh et al. 2016):

- **Measure or monitor (M)** data and information on emissions, mitigation actions, and support. This may entail direct physical measurement of GHG emissions, estimating emissions or emissions reductions utilizing activity data and emission factors, calculating changes relevant to sustainable development, and collecting information about support for climate change mitigation
- **Report (R)** by compiling this information in inventories and other standardized formats to make it accessible to a range of users and facilitate public disclosure of information.
- **Verify (V)** by periodically subjecting the reported information to some form of review or analysis or independent assessment to establish completeness and reliability. Verification helps to ensure accuracy and conformance with any established procedures, and can provide meaningful feedback for future improvement.

Furthermore, in Singh et al. (2016) study three types of mitigation-related MRV are discussed (Fig. 1.12):

- **MRV of GHG emissions**, conducted at national, organizational, and/or facility level to understand an entity's emissions profile and report it in the form of an emissions inventory.
- **MRV of mitigation actions** (e.g., policies and projects) to assess their GHG effects and sustainable development (non-GHG) effects as well as to monitor their implementation. This type of MRV focuses on estimating the change in GHG emissions or other non-GHG variables.
- **MRV of support** (e.g., climate finance, technology transfer, and capacity building) to track provision and receipt of climate support, monitor results achieved, and assess impact.

According to the MRV framework, currently, the reporting requirements and the timetable for the submission of national reports are different for Annex I Parties and Parties not included in Annex I to the Convention (non-Annex I Parties), in accordance with the principle of common but differentiated responsibilities and respective capabilities (Table 1.2).

However, most recently with the introduction of the Paris Agreement, it was agreed that all countries will provide emissions data and track progress against their contributions. MRV systems will be a significant component in effectively tracking and improving the implementation of mitigation goals and policies articulated under countries' Nationally Determined Contributions (NDCs). In order to build confidence in the process, the Paris Agreement includes a key element referred to as the **Enhanced Transparency Framework (ETF)** (Article 13, Annex, UNFCCC 2015) for action and support. Designed to promote transparency and mutual trust, the

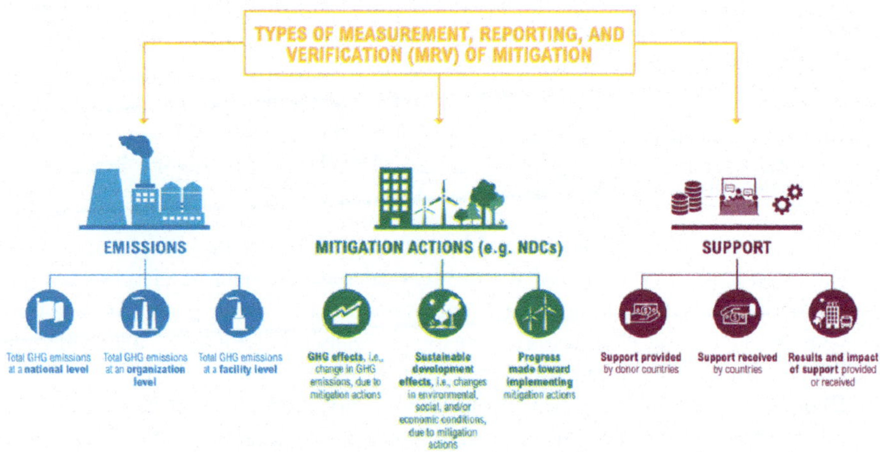

Source: (Singh et al., 2016)

Fig. 1.12 Types of mitigation related MRV
Source: Singh et al. (2016)

Table 1.2 Existing MRV arrangements for developed countries and developing countries

	Developed countries (Annex I Parties)	Developing countries (Non-Annex I Parties)
Reporting	Annual GHG inventory submissions	National communication
	National communication	Biennial update report
	Biennial report	GHG inventory as part of the national communication and biennial update report
	International Assessment and Review process (IAR)	International Consultation and Analysis process (ICA)
Technical review/ analysis	Technical review	Technical analysis of biennial update reports
	Review process of GHG inventories	
	Review reports of NCs and BRs	
Multilateral process	Multilateral assessment	Facilitative sharing of views

Source: https://unfccc.int/process-and-meetings/transparency-and-reporting/the-big-picture/what-is-transparency-and-reporting

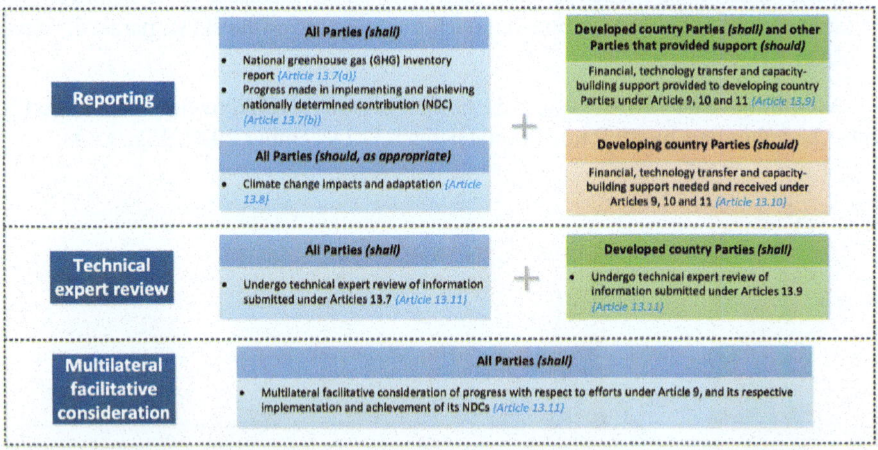

Source:https://unfccc.int/process-and-meetings/transparency-and-reporting/the-big-picture/what-is-transparency-and-reporting

Fig. 1.13 Article 13 of the Paris Agreement: transparency of action and support
Source: https://unfccc.int/process-and-meetings/transparency-and-reporting/the-big-picture/what-is-transparency-and-reporting

framework is based on existing transparency arrangements set up under the United Nations Convention on Climate Change (UNFCCC), commonly known as Measurement, Reporting and Verification (MRV) framework (Wartmann et al. 2018) (Fig. 1.13).

We are currently in a transitional phase in the shift from MRV framework toward the ETF. The ETF will supersede reporting of biennial reports (BRs) and biennial update reports (BURs) and the international assessment and review (IAR) and international consultation and analysis (ICA) processes for PA Parties (see decision 1/CP.24, para. 39):

- Reporting of the BR/BUR under the Convention will be superseded by reporting of the **biennial transparency report (BTR)** for PA Parties.
- Review of the BR and technical analysis (TA) of the BUR under the Convention will be superseded by **technical expert review** for PA Parties.
- Multilateral assessment (MA) and facilitative sharing of views (FSV) under the Convention will be superseded by the **facilitative multilateral consideration of progress (FMCP)** for PA Parties.

The following elements will continue to be reported under the Convention and are not superseded by the ETF:

- A **National Communications** (NC) must continue to be submitted by developed and developing countries.
- An **annual GHG inventory** must continue to be submitted by developed countries.
- The **technical annex on REDD+** (to be reported in BURs per decision 14/CP.19 (para. 7) for those Parties seeking results-based payments) is to be reported as an annex to the BTR and is reviewed during the review of the BTR (paras. 45–46 of 1/CP.24).

According to 18/CMA. 1, para. 3, the first **biennial transparency report (BTR)** must be submitted by all Parties no later than 31 December **2024** (Fig. 1.14).

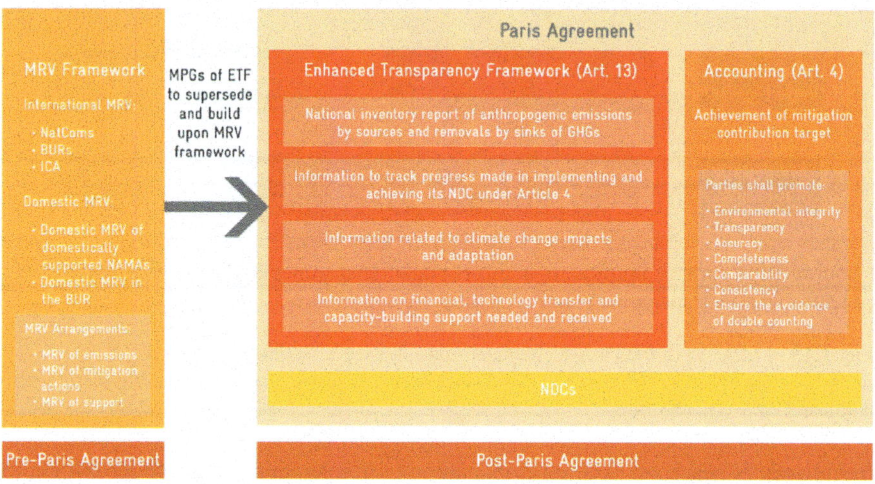

Source: (Wartmann et al., 2018)

Fig. 1.14 Transition from the MRV framework to the Enhanced Transparency Framework and accounting
Source: Wartmann et al. (2018)

1.3 Countries' Carbon Neutral Policies

In the light of international commitments and policies, worldwide, countries are committing to reach net zero emissions, adapting their national polices and adjusting their targets. A study by the Energy and Climate Intelligence Unit (2019) summarized the state-of-the-art of countries net zero commitments as they spread across the world (Fig. 1.15). As we can see from Fig. 1.15, Bhutan and Suriname are already carbon neutral and Norway and Sweden have already set their targets in national

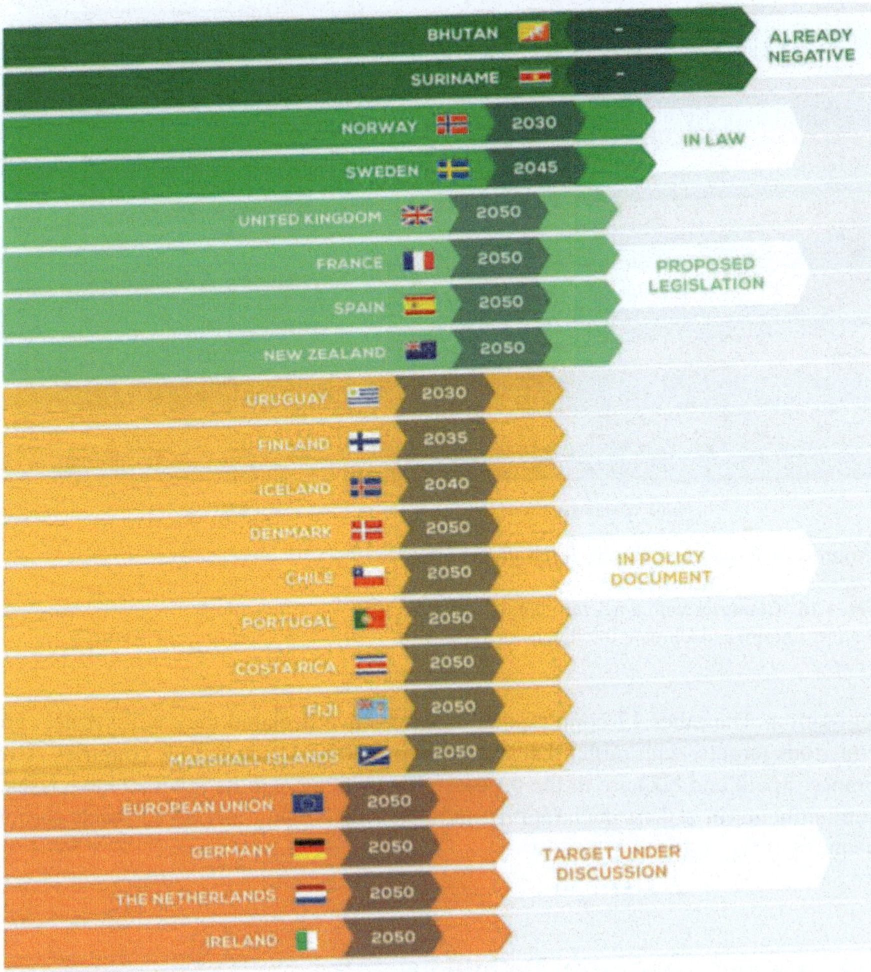

Source: (Energy and Climate Intelligence Unit, 2019)

Fig. 1.15 Countries' Carbon Neutral policies
Source: Energy and Climate Intelligence Unit (2019)

Country	Target date	Formality	Does the target cover all GHGs?	Does the target exclude international offsets?	Does the target include international aviation and shipping?	Does the target include interim targets?
Suriname	N/A	Achieved	N/A	N/A	N/A	N/A
Bhutan	N/A	Achieved	N/A	N/A	N/A	N/A
Norway	2030	In law	Yes	No	No	Yes
Uruguay	2030	In policy document	Yes	Unclear	No	No
Finland	2035	In policy document	Yes	Yes	No	No
Iceland	2040	In policy document	Yes	Yes	No	No
Sweden	2045	In law	Yes	No	No	Yes
Germany	2050	Target under discussion	Yes	Unclear	No	No
United Kingdom	2050	Proposed legislation	Yes	No	Yes	Yes
France	2050	Proposed legislation	Yes	Unclear	No	No
Spain	2050	Proposed legislation	Yes	Unclear	No	Yes
The Netherlands	2050	Target under discussion	Yes	Unclear	No	Yes
Ireland	2050	Target under discussion	Unclear	Unclear	Unclear	Yes
Denmark	2050	In policy document	Yes	Yes	No	No
Chile	2050	In policy document	Yes	Unclear	Unclear	Unclear
Portugal	2050	In policy document	Yes	Yes	No	No
New Zealand	2050	Proposed legislation	No	No	No	Yes
Costa Rica	2050	In policy document	Yes	Yes	No	No
Fiji	2050	In policy document	Yes	Unclear	No	Yes
Marshall Islands	2050	In policy document	Yes	Unclear	No	No

Table 1: Countries with a net zero commitment by 2050 Source: ECIU data

Source: (Energy and Climate Intelligence Unit, 2019)

Fig. 1.16 Countries with a net zero commitment by 2050
Source: Energy and Climate Intelligence Unit (2019)

legislation. The other 17 countries have set or have declared they will set net zero emissions targets, with target dates in or before 2050. Of these four countries (UK, France, Spain and NZ) are in the process of legislating and the others 9 have issued commitments in policy declarations and 4 are discussing targets to reach carbon neutrality (Fig. 1.16).

1.4 Carbon Neutrality Definitions

Instead of the rather technical term net zero emissions very often more colloquial terms as carbon neutral or climate neutral tend to be used. These two concepts are not identical. Moreover, due to a lack of specific, clear and well shared definition, they can be understood and used in a different way depending on the context (Emele et al. 2019).

- **Carbon neutral or Net zero carbon emissions** means that all activities lead to net zero CO2 emissions. This can be achieved by either reducing CO2 emissions to zero or compensating remaining CO2 emissions by sequestering the same amount of CO2 permanently from the atmosphere and depositing in a sink. While carbon neutrality implies net zero CO2 emissions there can still be significant amounts of non-CO2 emissions like methane (CH4), nitrous oxide (N2O) and fluorinated gases. Decarbonization is the process of reducing CO2 emissions and aiming for carbon neutrality.[1]
- In contrast the term **climate neutral or Net zero GHG emissions** covers not only CO2 but all greenhouse gases. Climate neutrality is achieved when the activities result in net zero GHG emissions. This does not mean that every single greenhouse gas has to reach net zero but the sum of all greenhouse gases. The different effects of greenhouse gases can be compared by the metric of global warming potentials (GWP), which is used to measure all emissions in "carbon equivalent".[2]

The differentiation between carbon neutral and climate neutral is remarkably relevant in modelling: many carbon models were developed to calculate only CO2 emissions while neglecting non-CO2 emissions. For example, energy models would typically treat bioenergy use as a zero-emission fuel, though depending on conditions, this process can generate significant emissions of very potent greenhouse gases (CH4 and N2O) or reinforce natural sinks. More generally, the concept of climate neutrality requires to address the much more complex accounting of life cycle emissions and removals, rather than a simpler measurement of gross CO2 emissions. Also, as this comprehensive approach is increasingly developed, the concept of carbon neutral tends to be used in the sense of a carbon equivalent neutrality, that is similar to climate neutrality (Emele et al. 2019) .

Moreover, different definitions of carbon neutrality can be found in literature. Table 1.3 summarize the most popular ones.

While the terms "carbon neutrality", "climate neutrality" and "net-zero emissions" have distinct definitions in the scientific context, these terms have been used with more flexibility and often inconsistency in the corporate sector (Pineda and Faria 2019). While in the corporate sector these terms are often used to indicate different approaches pursued by companies to mitigate their impact on the climate, in the scientific context, these terms are used distinctly to reflect the scope of climate forcers being neutralized. (Pineda and Faria 2019). As for example GHG emissions, CO2 or GHG emissions and other radiative forcing (Fig. 1.17).

---

[1] https://www.wri.org/blog/2015/12/cop21-glossary-terms-guiding-long-term-emissions-reduction-goal

[2] https://www.wri.org/blog/2015/12/cop21-glossary-terms-guiding-long-term-emissions-reduction-goal

Table 1.3 Carbon neutrality definitions

Concept	Focus	Definition	Source	Reference
Carbon neutrality	GHG	Condition in which during a specified period there has been no net increase in the global emission of **greenhouse gases** to the atmosphere as a result of the greenhouse gas emissions associated with the subject during the same period	PAS2060	BSI (2014)
Carbon neutrality	Carbon	Carbon neutrality means having a balance between emitting **carbon** and absorbing carbon from the atmosphere in carbon sinks.	European Parliament	European Parliament (2019)
Carbon neutral	Carbon emissions	Carbon neutral is a term used to describe the state of an entity (such as a company, service, product or event), where the **carbon emissions** caused by them have been balanced out by funding an equivalent amount of carbon savings elsewhere in the world.	Carbon footprint	Carbon Footprint Ltd. (2019)
Carbon neutrality	Carbon emissions	Carbon neutrality, or having a net zero carbon footprint, refers to achieving net zero **carbon emissions** by balancing a measured amount of carbon released with an equivalent amount sequestered or offset.	UNEP	United Nations Environment Programme (UNEP) (2019)
Carbon neutrality/net zero co2 emissions	Co2	Net zero carbon dioxide (CO2) emissions are achieved when anthropogenic **CO2 emissions** are balanced globally by anthropogenic CO2 removals over a specified period. Net zero CO2 emissions are also referred to as carbon neutrality.	IPCC	IPCC (2018b)
Net zero emissions	GHG	Net zero emissions are achieved when anthropogenic emissions of **greenhouse gases** to the atmosphere are balanced by anthropogenic removals over a specified period.	IPCC	IPCC (2018b)
Carbon neutral	GHG	Achieve a balance between anthropogenic emissions by		United Nations Framework

(continued)

Table 1.3 (continued)

Concept	Focus	Definition	Source	Reference
		sources and removals by sinks of **GHGs** in the second half of the century.	Paris Agreement art. 4	Convention on Climate Change (2015)
Climate neutrality	Emissions	Climate neutrality is achieved by balancing the amount of **emissions** generated with the Earth's natural capacity of to absorb them. Some use the term "carbon neutrality" or "net zero" as synonyms of climate neutrality.	UNFCC	UNFCC (2019b)

Source: Author's own elaboration

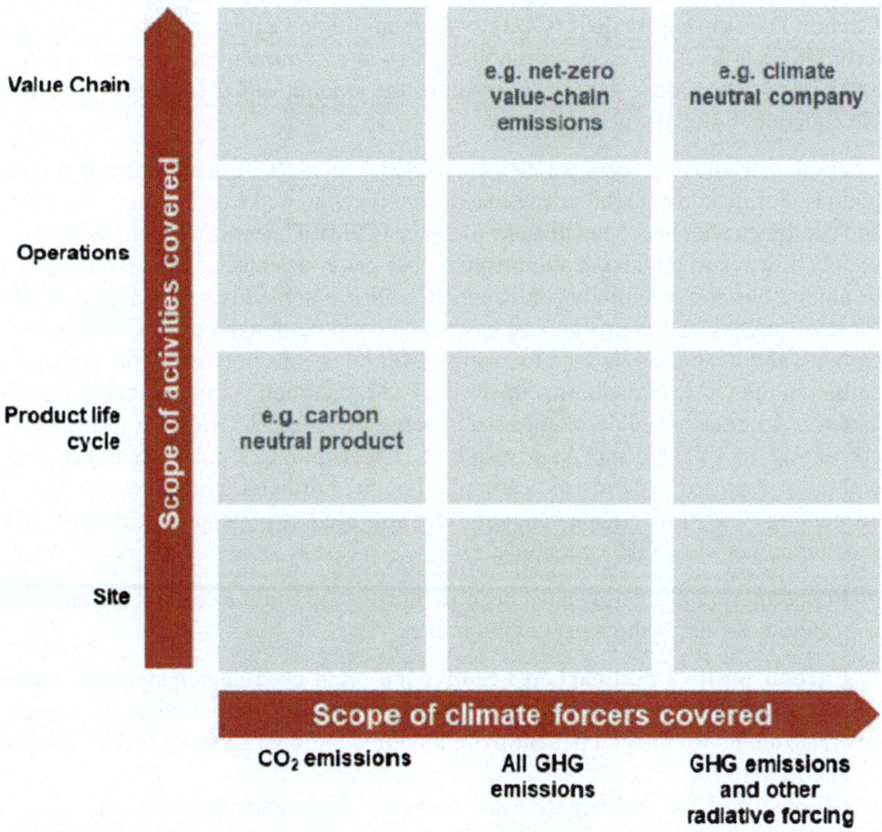

Source: (Pineda and Faria, 2019).

Fig. 1.17 Boundary of corporate climate neutrality targets
Source: Pineda and Faria (2019)

This issue is also related to the misleading use of the terminology used to describe GHG emissions (Watkins and Durning 2012). In this sense Brander (2012) provided some definitions:

Greenhouse Gases A greenhouse gas (or GHG for short) is any gas in the atmosphere which absorbs and re-emits heat, and thereby keeps the planet's atmosphere warmer than it otherwise would be. The main GHGs in the Earth's atmosphere are water vapor, carbon dioxide (CO2), methane (CH4), nitrous oxide (N2O) and ozone. Often these GHGs are referred to as the "Kyoto gases".

Carbon Dioxide Carbon dioxide (CO2) is the most common GHG emitted by human activities, in terms of the quantity released and the total impact on global warming. As a result, the term "CO2" is sometimes used as a shorthand expression for all greenhouse gases, however, this can cause confusion, and a more accurate way of referring to a number of GHGs collectively is to use the term "carbon dioxide equivalent" or "CO2e" (explained below).

Carbon Dioxide Equivalent (CO2e) "Carbon dioxide equivalent" or "CO2e" is a term for describing different greenhouse gases in a common unit. For any quantity and type of greenhouse gas, CO2e signifies the amount of CO2 which would have the equivalent global warming impact.

Carbon Carbon is a chemical element which is present in many gases and compounds. For example, carbon combines with oxygen to make carbon dioxide (CO2), and combines with hydrogen to make methane (CH4). The term "carbon" is used in a variety of ways when talking about greenhouse gas emissions, and therefore tends to be ambiguous and potentially confusing. "Carbon" is sometimes used as a shorthand for referring to CO2, or greenhouse gases in general (although not all GHGs contain carbon), and it can also be used to express CO2 emissions in terms of the amount of carbon in the CO2. In addition, converting CO2 to carbon is not particularly useful as doing so does not allow comparisons between different GHGs, in the way that converting to CO2e does. As a result, it is less and less common to see CO2 emissions reported in terms of "carbon", though shorthand terms such as "carbon accounting" and "low carbon economy" are still used as popular proxies for "GHG accounting" or "low GHG economy".

Furthermore, other terms have been frequently used in the carbon neutral narrative. Below we listed the most common ones.

- **Carbon positive and carbon negative** are often used interchangeably; means that an activity goes beyond achieving net zero carbon emissions to actually create an environmental benefit by removing additional carbon dioxide from the atmosphere.
- **Climate positive and negative** are used interchangeably; means that an activity goes beyond achieving net zero GHG emissions to actually create an environmental benefit by removing additional GHG emissions from the atmosphere
- **Decarbonization** The process by which countries, individuals or other entities aim to achieve zero fossil carbon existence. Typically refers to a reduction of the carbon emissions associated with electricity, industry and transport (IPCC 2018b)

- **Low-carbon (or low-emissions) transformation** conveys intent to transform our systems (e.g. technology, energy, political, infrastructure) and limit warming to 2 °C. (WRI 2015)
- **Carbon budget** This term refers to three concepts in the literature: (1) an assessment of carbon cycle sources and sinks on a global level, through the synthesis of evidence for fossil fuel and cement emissions, land-use change emissions, ocean and land CO2 sinks, and the resulting atmospheric CO2 growth rate. This is referred to as the global carbon budget; (2) the estimated cumulative amount of global carbon dioxide emissions that that is estimated to limit global surface temperature to a given level above a reference period, taking into account global surface temperature contributions of other GHGs and climate forcers; (3) the distribution of the carbon budget defined under (2) to the regional, national, or sub-national level based on considerations of equity, costs or efficiency (IPCC 2018b) (Fig. 1.18).

However, besides the terminology confusion, it is commonly accepted that when talking of Carbon neutrality, we refer to all the GHG emissions as defined by the IPCC guidelines issued in 2006 and revised in 2019.[3]

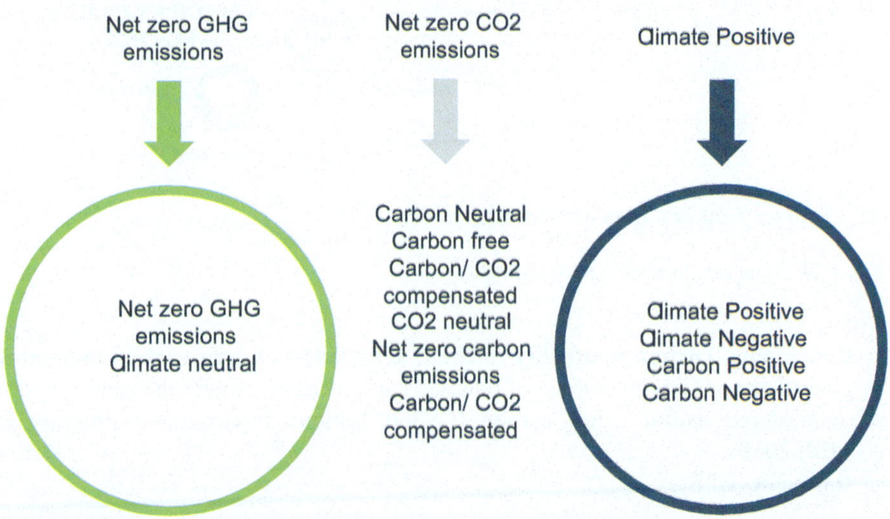

Source: Authors' own elaboration.

Fig. 1.18 Carbon neutral definitions

[3]The GHG Protocol requires inclusion of all the GHGs required by the UNFCCC/Kyoto Protocol at the time a corporate or product inventory is being compiled. These GHGs are currently: carbon dioxide (CO 2), methane (CH 4), nitrous oxide (N 2 O), hydrofluorocarbons (HFCs), perfluorocarbons (PFCs), sulphur hexafluoride (SF 6), and nitrogen trifluoride (NF 3).

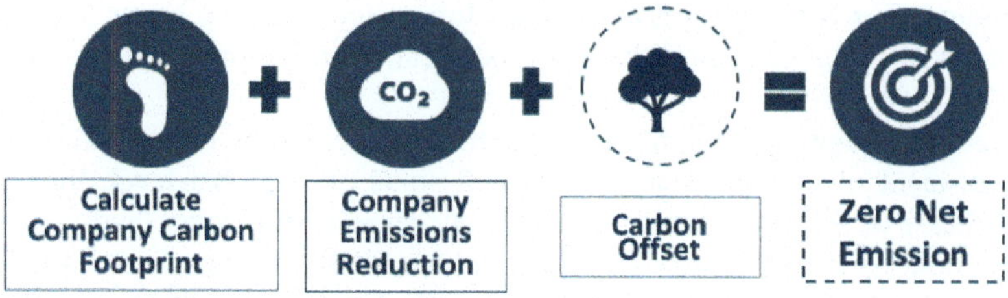

Source: Authors' own elaboration

Fig. 1.19 Main step to reach carbon neutrality

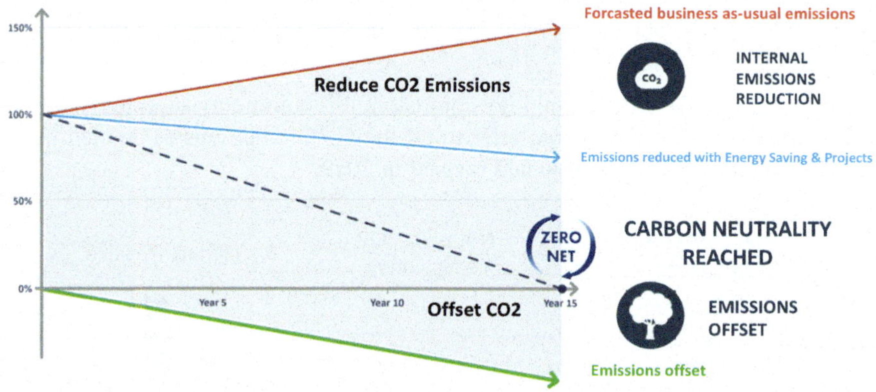

Source: Authors' own elaboration

Fig. 1.20 Main step to reach carbon neutrality

Concluding, **carbon neutrality** refers to **achieving net zero carbon emissions** by balancing a measured amount of released carbon emission with the same amount of sequestered and/or offset carbon (United Nations Environment Programme (UNEP) 2019).

It is achieved by:

- Reducing the company's carbon intensity by improving processes and technologies (BSI 2014)
- Offset emissions by financially supporting third-party certified offsetting projects (BSI 2014) (Figs. 1.19 and 1.20)

References

Barilla Center for Food and Nutrition (2016) DoUBLE PYRAMID 2016 Eat better Eat less Food for all

Bedsted B, Mathieu Y, Leyrit C (2015) World wide views on climate and energy

Benin Ministère du Cadre de vie et du Développement Durable (2016) Stratégie de développement à faible intensité de carbone et résilient aux changements climatiques

Brander M (2012) Greenhouse gases, CO2, CO2e, and carbon: what do all these terms mean? Ecometrica

Braun R (2010) Social participation and climate change. Environ Dev Sustain 12:777–806. https://doi.org/10.1007/s10668-009-9224-1

BSI (2014) PAS 2060:2014 Specification for the demonstration of carbon neutrality

Carbon Footprint Ltd (2019) What is carbon neutrality? [WWW document]. https://www.carbonfootprint.com/carbonneutrality.html

Directorate-General for Climate Action -European Commission (2019) Going climate-neutral by 2050

Emele, L., Marignac, Y., Petrovic, S., 2019. Modelling net zero emissions. Clim. Recon 2050 dialogues pathways policy

Energy and Climate Intelligence Unit (2019) Count down to zero:Plotting progress towards delivering net zero emissions by 2050

European Commission (2017) Attitudes of European citizens towards the environment. Report. Special Eurobarometer 468, Special Eurobarometer 468

European Commission (2018) A clean planet for all. A European strategic long-term vision for a prosperous, modern, competitive and climate neutral economy – Communication from the Commission to the European Parliament, the Council, the European and Social Committee and the Committe

European Commission- Directorate General (2019) Special Eurobarometer 490 "Climate Change" – report. https://doi.org/10.2834/00469

European Parliament (2019) What is carbon neutrality and how can it be achieved by 2050? [WWW document]

Federal Ministry for the Environment Nature Conservation Building and Nuclear Safety (BMUB) (2016) Climate Action Plan 2050: Principles and goals of the German government's climate policy 6

Feucht Y, Zander K (2018) Consumers' preferences for carbon labels and the underlying reasoning. A mixed methods approach in 6 European countries. J Clean Prod 178:740–748. https://doi.org/10.1016/j.jclepro.2017.12.236

Government of Canada -Environment and Climate Change Canada (2016) Canada's mid-century long-term low-greenhouse gas development strategy

Government of Fiji (2018) Fiji Low Emission Development Strategy 2018–2050

Government of Japan (2019) The Long-term Strategy under the Paris Agreement

Government of Mexico (2016) Mexico's Climate Change Mid-Century Strategy 100

Intergovernmental Panel on Climate Change (2014) Climate Change 2014 Mitigation of Climate Change. https://doi.org/10.1017/CBO9781107415416

IPCC (2018a) Summary for policymakers, global warming of 1.5 °C. an IPCC special report on the impacts of global warming of 1.5°C above pre-industrial levels and related global greenhouse gas emission pathways, in the context of strengthening the global response to the threat of climate change

IPCC (2018b) Annex I: glossary, in: global warming of 1.5°C. an IPCC special report on the impacts of global warming of 1.5 °C above pre-industrial levels and related global greenhouse gas emission pathways, in the Context of Strengthening the Global Response to the Threat of Climate Change

Kelly L, Franse T (2019) Climate action for today and tomorrow: the relationship between NDCs and LTSs [WWW document]. World Resources Institute

Masson-Delmotte V, Zhai P, Pörtner HO, Roberts D, Skea J, Shukla PR, Pirani A, Moufouma-Okia W, Péan C, Pidcock R, Connors S, Matthews JBR, Chen Y, Zhou X, Gomis MI, Lonnoy E, Maycock T, Tignor M, Waterfield T (2018) Global warming of 1.5 °C, report of the Intergovernmental Panel on Climate Change

Ministère de la Transition écologique et solidaire (2015) La stratégie nationale bas-carbone (SNBC)

Ministry of the Environment of the Czech Republic (2017) Climate protection policy of the czech republic: executive summary, Revenue Recognition: Understanding and Implementing the New Standard. https://doi.org/978-80-7212-626-2

Peattie K, Peattie S, Ponting C (2009) Climate change: a social and commercial marketing communications challenge. EuroMed J Bus 4:270–286. https://doi.org/10.1108/14502190910992693

Pineda AC, Faria P (2019) Towards a science-based approach to climate neutrality in the corporate sector discussion paper. Sci Based Targets:1–29

Portuguese Ministry of Environment and Energy Transition (2019) Long-term strategy for carbon neutrality of the Portuguese Economy by 2050 – Roadmap to Carbon Neutrality 2050 R 262/2019

Rocha M, Falduto C (2019) Key questions guiding the process of setting up long-term low-emissions development strategies. OECD

Singh N, Finnegan J, Levin K (2016) MRV 101: understanding measurement, reporting, and verification of climate change mitigation, Washington, DC

Sirieix L, Delanchy M, Remaud H, Zepeda L, Gurviez P (2013) Consumers' perceptions of individual and combined sustainable food labels: a UK pilot investigation. Int J Consum Stud 37:143–151. https://doi.org/10.1111/j.1470-6431.2012.01109.x

The Republic of the Marshall Islands (2018) 2050 Climate Strategy "Lighting the way". https://doi.org/.1037//0033-2909.I26.1.78

The White House (2016) United States Mid-Century Strategy for Deep Decarbonization 111

Thøgersen J, Nielsen KS (2016) A better carbon footprint label. J Clean Prod 125:86–94. https://doi.org/10.1016/j.jclepro.2016.03.098

UK Department for Business Energy & Industical Strategy; HM Government (2017) The clean growth strategy: leading the way to a low carbon future, UK Gov.

Ukraine Ecology and Natural Resources Ministry (2018) Ukraine 2050 low emission development strategy

UNFCC (2019a) Communication of long-term strategies [WWW document]. URL https://unfccc.int/process/the-paris-agreement/long-term-strategies. Accessed 8.26.19

UNFCC (2019b) What is climate neutrality? [WWW document]. https://unfccc.int/climate-action/climate-neutral-now/resources-and-frequently-asked-questions#eq-1

UNFCCC (1992) Report of the intergovernmental negotiating Committee for a Framework Convention on Climate Change on the work of the second part of its fifth session, held at New York from 30 April to 9 May 1992. Addendum: A/AC.237/18 (Part II)/Add.1 [WWW document]. URL http://unfccc.int/resource/docs/a/18p2a01.pdf

UNFCCC (2008) Report of the conference of the parties on its thirteenth session, held in Bali from 3 to 15 December 2007. http://Unfccc.Int/Resource/Docs/2007/Cop13/Eng/06a01.Pdf. https://doi.org/10.1016/j.biocon.2006.08.013

United Nations (2014) Climate Change Summit Not about Talk, but Action, Secretary-General Says at Press Conference, Stressing 'All of Us Can Make a Difference' [WWW Document]. https://www.un.org/press/en/2014/sgsm16186.doc.htm

United Nations Environment Programme (UNEP) (2014) UN Says Global Carbon Neutrality Should be Reached by Second Half of Century, Demonstrates Pathways to Stay Under 2°C Limit [WWW Document]

United Nations Environment Programme (UNEP) (2019) UN environment "walks the talk" on carbon neutrality [WWW document]. https://www.unenvironment.org/news-and-stories/story/un-environment-walks-talk-carbon-neutrality

United Nations Framework Convention on Climate Change (2015) PARIS AGREEMENT – conference of the parties COP 21. Adopt. Paris Agreement. Propos. by Pres. https://doi.org/FCCC/CP/2015/L.9/Rev.1

Vallejo L, Rankovic A, Colombier M, Iddri JV (2018) Carbon neutrality: taking on the global challenge for ambitious climate action. IDDRI

Vanclay JK, Shortiss J, Aulsebrook S, Gillespie AM, Howell BC, Johanni R, Maher MJ, Mitchell KM, Stewart MD, Yates J (2011) Customer response to carbon labelling of groceries. J Consum Policy 34:153–160. https://doi.org/10.1007/s10603-010-9140-7

Wartmann S, Raúl Salas R, Blank D (2018) Deciphering MRV, accounting and transparency for the post-Paris era

Watkins J, Durning B (2012) Carbon definitions and typologies in environmental impact assessment: greenhouse gas confusion? Impact Assess Proj Apprais. https://doi.org/10.1080/14615517.2012.730224

WRI (2019) Enhancing NDCS: a guide to strengthening national climate plans by 2020

Chapter 2
Carbon Neutrality in the Agri-Food Sector

The chapter introduces the dual relationship between climate change and the agri-food sector. While agriculture is one of the main sources of GHG emissions, this sector is an inexhaustible source of solutions related to mitigation and adaptation to climate change.

2.1 Agri-Food Systems Are Facing a Growing Climate Change Problem and Opportunity

In this context, the importance of low carbon and climate neutral products is critical, especially for the agri-food sector. Agri-food is responsible for excessive resource use and environmental damage (Shukla et al. 2019):

- Contributes between 21% and 37% of GHG emissions;
- Food waste is responsible for 8% of GHG emission;
- Contributes more than 58% of nitrogen emissions (300x more harmful of Co2);
- Is a major driver of the loss of biodiversity;
- Uses about 1/3 of the planet's land for food production;
- Accounts for 70% of freshwater withdrawals.

The impact of agri-food also incurs huge social costs:

- By 2050, with the population nearing 10 billion, food production would to rise 70%;
- 1.3 billion tons of food are wasted every year and 820 million people go hungry;
- Effects of Climate Change will hit hard in developing countries where large numbers of people are employed in agriculture as land productivity decrease;

Agriculture alone (including land use change) is responsible for nearly 14–24% of global anthropogenic emissions (Vermeulen et al. 2012). Agriculture (excluding

A. Acampora et al., *Carbon Neutrality in the Agri-food Sector*,
https://doi.org/10.1007/978-3-030-88048-4_2

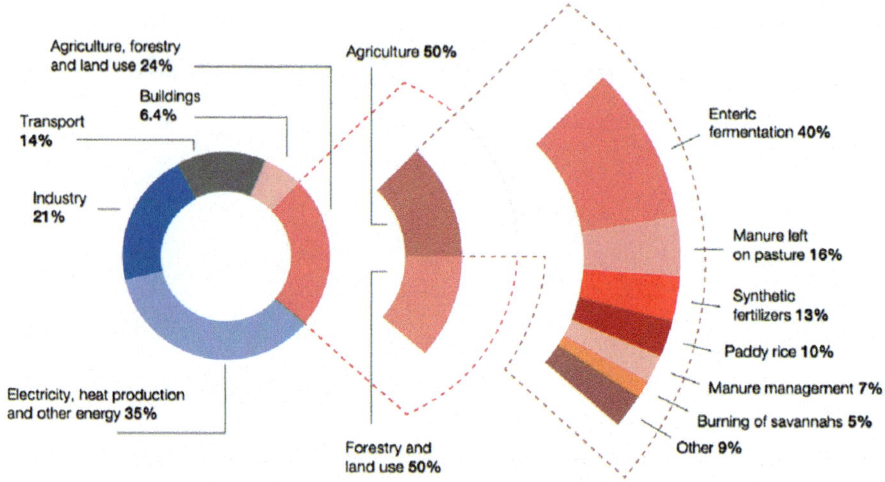

Source: (Richards et al., 2019)

Fig. 2.1 Agriculture and Forestry and land use GHG emissions
Source: Richards et al. (2019)

forestry and other land uses) contributes approximately 12 per cent of global GHG emissions. Methane (CH4) and nitrous oxide (N2O) are the primary GHGs produced by agricultural activities (Richards et al. 2019). There are four primary sources of GHG emissions from agriculture: enteric fermentation, livestock manures, fertilizers and paddy rice (Fig. 2.1).

Today's food supply chain creates ~13.7 billion metric tons of carbon dioxide equivalents (CO2eq), 26% of anthropogenic GHG emissions of which 31% come from livestock and fisheries, 27% from crop production, 24% from land use and 18% from the supply-chain (e.g. Retail (3%), Packaging (5%), Transport (6%) and Food processing (4%))(Fig. 2.2)(Poore and Nemecek 2018).

As we can see from Fig. 2.3 of 26% of GHG emissions related to food the 58% derive from animal products with beef and lamb contributing to half of these emissions (Fig. 2.3).

However, Poore and Nemecek (2018) found great variability in impact among both products and producers. This variability, even among producers in similar geographic regions, implies substantial potential to reduce environmental impacts and enhance productivity in the food system. For many products, impacts are skewed by producers with particularly high impacts. This creates opportunities for targeted mitigation, making an immense problem more manageable. For example, for beef originating from beef herds, the highest-impact 25% of producers represent 56% of the beef herd's GHG emissions. For example, beef cattle raised on deforested land is responsible for 12 times more greenhouse gas emissions than cows reared on natural pastures (Fig. 2.4).

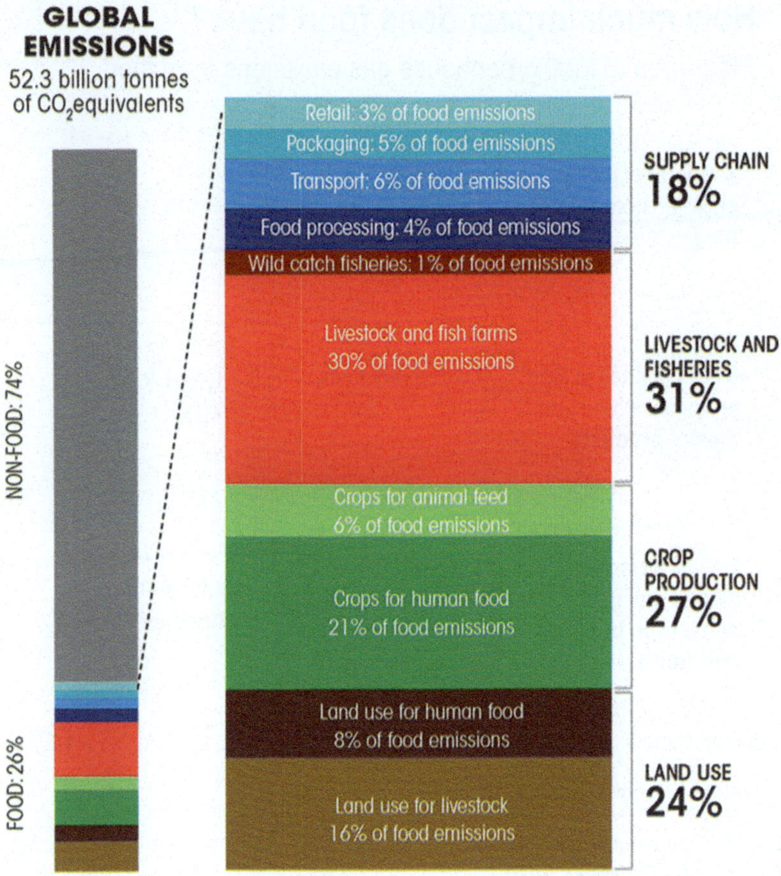

Fig. 2.2 GHG emissions from food production
Source: Poore and Nemecek (2018)

2.2 Agriculture as Source of Solutions in Fighting Climate Change: Turning a Planetary Pollutant into an Asset

Agriculture is at the same time the main cause and the main victim of the ongoing environmental crisis. Agricultural practices, in fact, produce significant volumes of GHG emission, the main cause of climate change. Nevertheless, this sector suffers the most negative impacts of climate change, in terms of reduced productivity and increased risks related to food safety. Experts argue that even low levels of global warming will reduce agricultural production and trigger greater variability in production in lower latitude regions of the world. The negative effects on agricultural production will be exacerbated by more frequent extreme weather events (such as floods, heat waves and droughts) (European Commission 2015). Small farmers and

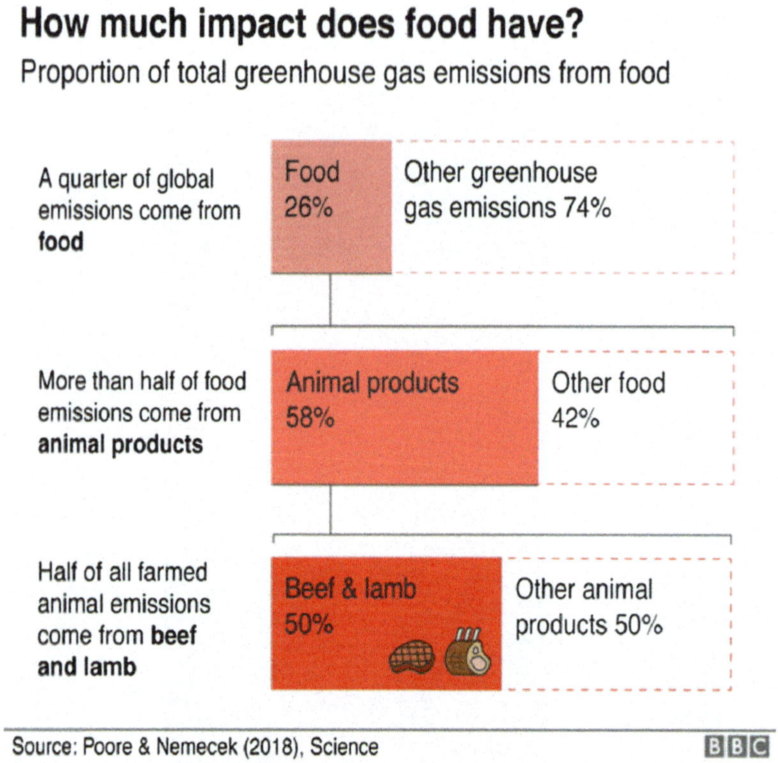

Source: (Poore and Nemecek, 2018)

Fig. 2.3 GHG emissions from food
Source: Poore and Nemecek (2018)

subsistence farmers will be particularly affected as they are less able to adapt. While some of the expected consequences could be beneficial for agriculture in some European regions, particularly the northern areas (lengthening of the growing season and improvements in agricultural production due to milder climatic conditions), it is expected that most of the consequences will be negative, will bring economic losses and will occur in regions already under pressure due to socio-economic and other factors, such as water scarcity (European Commission 2015). However, agriculture has a great potential in the fight against climate change, as well as reducing GHG emissions, through the introduction of sustainable agricultural practices, it is the only sector that has the capacity to remove GHGs safely and cost-effectively from the atmosphere without reducing productivity (Mayhew 2016). Moreover, the agricultural sector has the potential to benefit from synergies between climate change adaptation and mitigation (Altieri and Nicholls 2017). A recent Intergovernmental Panel on Climate Change (IPCC) Special Report on Global Warming of 1.5 °C confirms there is an **important role for land use sectors in stabilizing global temperatures** (IPCC 2018).

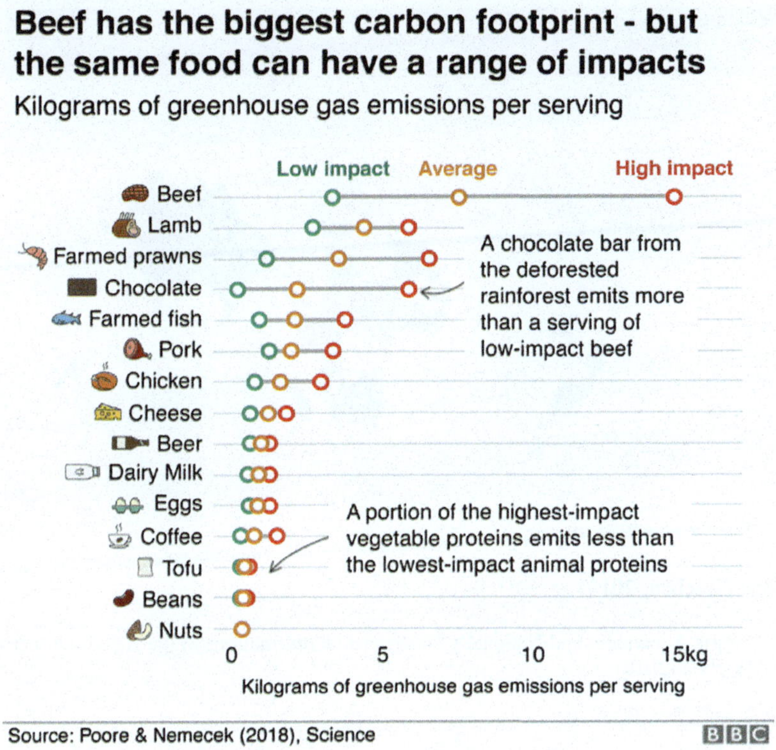

Source: (Poore and Nemecek, 2018).

Fig. 2.4 GHG emissions from different foods
Source: Poore and Nemecek (2018)

In this sense, in 2019, FAO developed a report performing a global assessment on how food value chains and the agri-food sector have been considered in the NDCs (Wieben 2019). From the report emerged that Out of the 163 submissions analyzed, over half of them, 55%, provide references to various aspects of food value chains. Figure 2.5 provides a geographical overview of the countries, which have cited food value chain aspects in their NDCs. All of them, with the exception of one country, are Non-Annex I Parties to the Convention.

All Parties' NDCs include contributions related to climate change mitigation, through the reduction or avoidance of anthropogenic GHG emissions covered by the UNFCCC as well as through increased sinks such as forest cover. Forty-five NDCs refer to specific mitigation measures, including approaches to retain value and minimize waste from the agri-food sector.

Most frequently cited are measures for managing and utilizing biomass waste, mainly as a source of renewable energy or as organic fertilizer. The majority of the NDCs that provide relevant references as part of their mitigation measures are

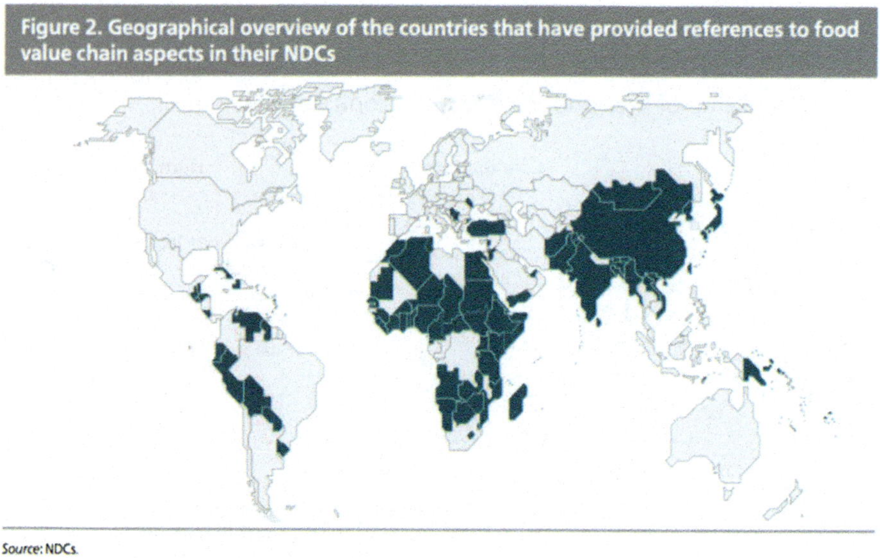

Source: (Wieben, 2019)

Fig. 2.5 Paris Agreement and Nationally Determined Contributions in the Agri-food sector
Source: Wieben (2019)

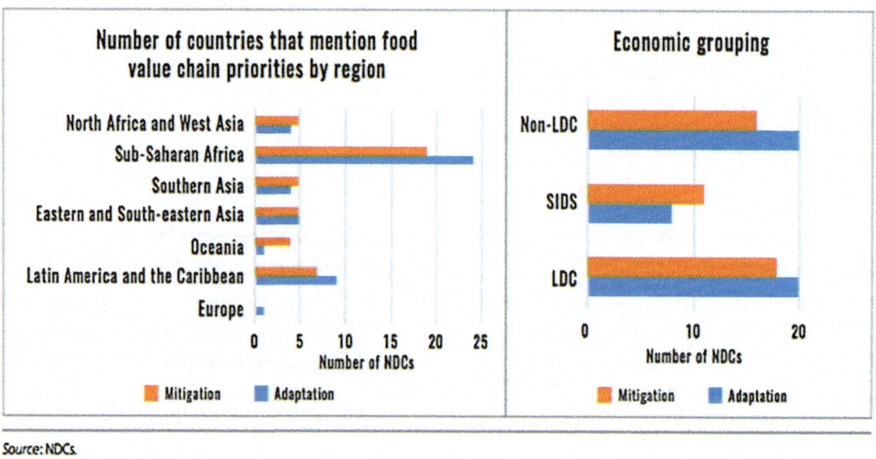

Source: (Wieben, 2019)

Fig. 2.6 Paris Agreement and NDCs in the Agri-food sector - Mitigation and adaptation measures
Source: Wieben (2019)

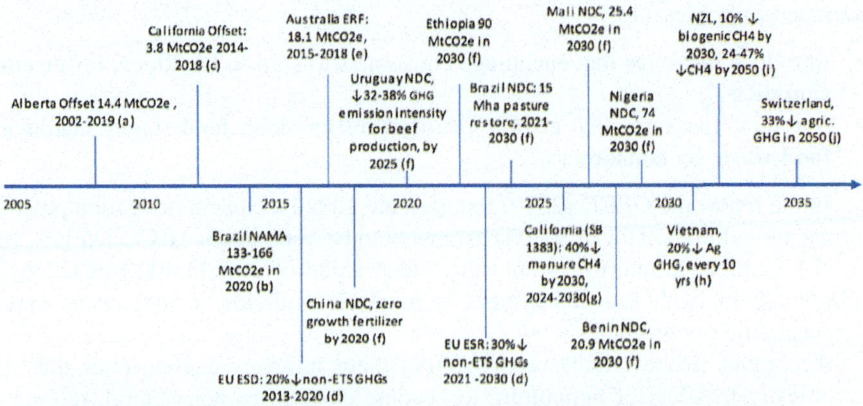

Note: Includes a selection of mitigation policies, national mitigation targets, and realised reductions. Within the EU, a number of countries have set specific targets and budgets for agriculture to contribute to Effort Sharing goals, which are elaborated for some countries in the text. An attempt was made to place the policies/targets in the midpoints of their timeframes.
Sources: (a) AEOR (2019); (b) Gebara and Thuault (2013); (c) CARB (2019); (d) European Commission (2019a); (e) Clean Energy Regulator (2019); (f) UNFCCC (2019); (g) CARB (2017); (h) Government of Vietnam (2019); (i) Ministry for the Environment, 2019; (j) Bundesamt für Landwirtschaft (2019).

Source: (OECD, 2019)

Fig. 2.7 A snapshot of mitigation policy progress in the agriculture sector
Source: OECD (2019)

predominantly from Sub-Saharan African Parties, and out of the 45 identified NDCs, 64% are either LDCs or SIDS (Fig. 2.6) (Wieben 2019).

Moreover, in a recent report by OECD (2019), a list of the main mitigation policy and target in the agricultural sector was elaborated. In the report the OECD reviewed both NDCs and the countries policies that specifically target the agriculture sector with explicit incentives or establishing a price on emissions or funding mitigation measures in agriculture (Fig. 2.7).

According to a recent report elaborated by the OECD (2019) four broad options could be implemented in the agriculture sector to mitigate GHG emissions. The first two encompass **supply-side measures** and the latter two cover **demand-side measures** (OECD 2019):

Supply-Side Measures:

- Introduce **farm practices** that reduce agricultural non-carbon dioxide (non-CO2) emissions; including methane (CH4) and nitrous oxide (N2O).
- Introduce **practices to remove CO2 from the atmosphere and accumulate as carbon in vegetation and soils**, or that reduce emissions from the degradation and removal of these carbon stocks

Demand-Side Measure:

- Introduce measures that encourage consumers to shift to **healthier, lower emission diets.**
- Introduce measures that **reduce product losses** along food supply chains and **food waste by consumers**.

In the report the OECD (2019) reported the global technical mitigation potential of the agricultural sector in 2030 is estimated to be **5500–6000 MtCO2eq yr1**, with a 95% confidence interval around this mean value of 300–11,400 MtCO2eq yr1 (Smith 2012), with full deployment of available emission reduction and carbon sequestration opportunities.

According to the OECD report (2019) these numbers demonstrates that it is technically feasible for agriculture to become close to carbon neutral, relying on supply-side mitigation measures alone, although this depends on optimistic assumptions about the potential of soil carbon sequestration (SCS).

The most recent Assessment Report of the IPPC report (IPCC 2018), based on results from several studies (Golub et al. 2009; Nauclér and Enkvist 2009; Rose et al. 2012; Smith et al. 2008), found that emission reductions for agriculture of 0.03–2.6 GtCO2eq are possible at USD 50 tCO2eq-1, and 0.2–4.6 GtCO2eq at USD 100 tCO2eq-1 in 2030 (Smith et al. 2014). This wide range reflects the different coverage of mitigation sources and methodologies used. For example, the higher figures in these ranges include SCS measures.

The global potential for SCS is estimated to be large but highly uncertain. Smith (2016) reports the mean global potential for SCS in agricultural soils as 1.5 Gt CO2 yr-1 and 2.6 GtCO2 yr1, at carbon prices of USD 20/tCO2eq and USD 100/tCO2eq, respectively.

In addition, the mitigation potential from increasing carbon stocks in vegetation is substantial. For example, Golub et al. (2013) found that a policy package that combined a tax on agricultural emissions and a subsidy to sequester carbon in forest biomass at USD 27/tCO2eq could mitigate 5.3 GtCO2eq yr-1 of emissions. Most of this potential was attributed to the sequestration of carbon in forest biomass, particularly from avoided deforestation and the conversion of forest to agricultural land (i.e. from changes in the extensive margin between forestry and agriculture) (OECD 2019).

Furthermore, the IPCC released in 2019 a special Report on Climate Change and Land on climate change, desertification, land degradation, sustainable land management, food security, and greenhouse gas fluxes in terrestrial ecosystems (Shukla et al. 2019).

In the report the IPCC reviewed different mitigation response options that according to the the Panel have technical potential for >3 GtCO -eq $yr.^{-1}$ by 2050 through reduced emissions and Carbon Dioxide Removal (CDR) (high confidence). However, some of these options compete for land and other resources, while others may reduce the demand for land (high confidence). Moreover, the IPCC stated that estimates of the technical potential of individual response options sows that these are not necessarily additive.

The report shows that the largest potential for reducing AFOLU emissions are through reduced deforestation and forest degradation (0.4–5.8 GtCO -eq yr–1) (high confidence), a shift towards plant- based diets (0.7–8.0 GtCO -eq yr–1) (high confidence) and reduced food and agricultural waste (0.8–4.5 CO2-eq yr–1) (high confidence). Furthermore, the IPCC report stated that agriculture measures combined could mitigate 0.3–3.4 GtCO -eq yr–1 (medium confidence).

The options with largest potential for CDR are afforestation/reforestation (0.5–10.1 CO -eq yr–1) (medium confidence), soil carbon sequestration in croplands and grasslands (0.4–8.6 CO -eq yr–1) (high confidence) and Bioenergy with Carbon Capture and Storage (BECCS) (0.4–11.3 CO -eq yr–1) (medium confidence). While some estimates include sustainability and cost considerations, most do not include socio-economic barriers, the impacts of future climate change or non-GHG climate forcings.

In Fig. 2.8 mitigation potentials reflect the full range of low to high estimates from studies published after 2010, differentiated according to technical (possible with current technologies), economic (possible given economic constraints) and sustainable potential (technical or economic potential constrained by sustainability considerations). Medians are calculated across all potentials in categories with more than four data points. IPCC only include references that explicitly provide mitigation potential estimates in CO2-eq yr–1 (or a similar derivative) by 2050. Not all options for land management potentials are additive, as some may compete for land. Estimates reflect a range of methodologies (including definitions, global warming potentials and time horizons) that may not be directly comparable or additive. Results from IAMs are shown to compare with single option 'bottom-up' estimates, in available categories from the 2 °C and 1.5 °C scenarios in the SSP Database (version 2.0). The models reflect land management changes, yet in some instances, can also reflect demand-side effects from carbon prices, so may not be defined exclusively as 'supply-side'.

Furthermore, the IPCC (2019) report assessed different response options across adaptation, mitigation, combating desertification and land degradation, food security and sustainable development, and a select set of options deliver across all of these challenges. These options include, but are not limited to, sustainable food production, improved and sustainable forest management, soil organic carbon management, ecosystem conservation and land restoration, reduced deforestation and degradation, and reduced food loss and waste. Most of these response options have the potential to provide multiple co-benefits and can positively contribute to sustainable development. Particularly, according to this report the total technical mitigation potential from crop and livestock activities, and agroforestry is estimated as 2.3–9.6 GtCO eq yr–1 by 2050. The total technical mitigation potential of dietary changes is estimated as 0.7–8 GtCO eq yr–1 by 2050. Future land use depends, in part, on the desired climate outcome and the portfolio of response options deployed. All assessed modelled pathways that limit warming to 1.5 °C or well below 2 °C require land-based mitigation and land-use change, with most including different combinations of reforestation, afforestation, reduced deforestation, and bioenergy. A small number of modelled pathways achieve 1.5 °C with reduced land conversion and thus reduced consequences for desertification, land degradation, and food security (Fig. 2.9).

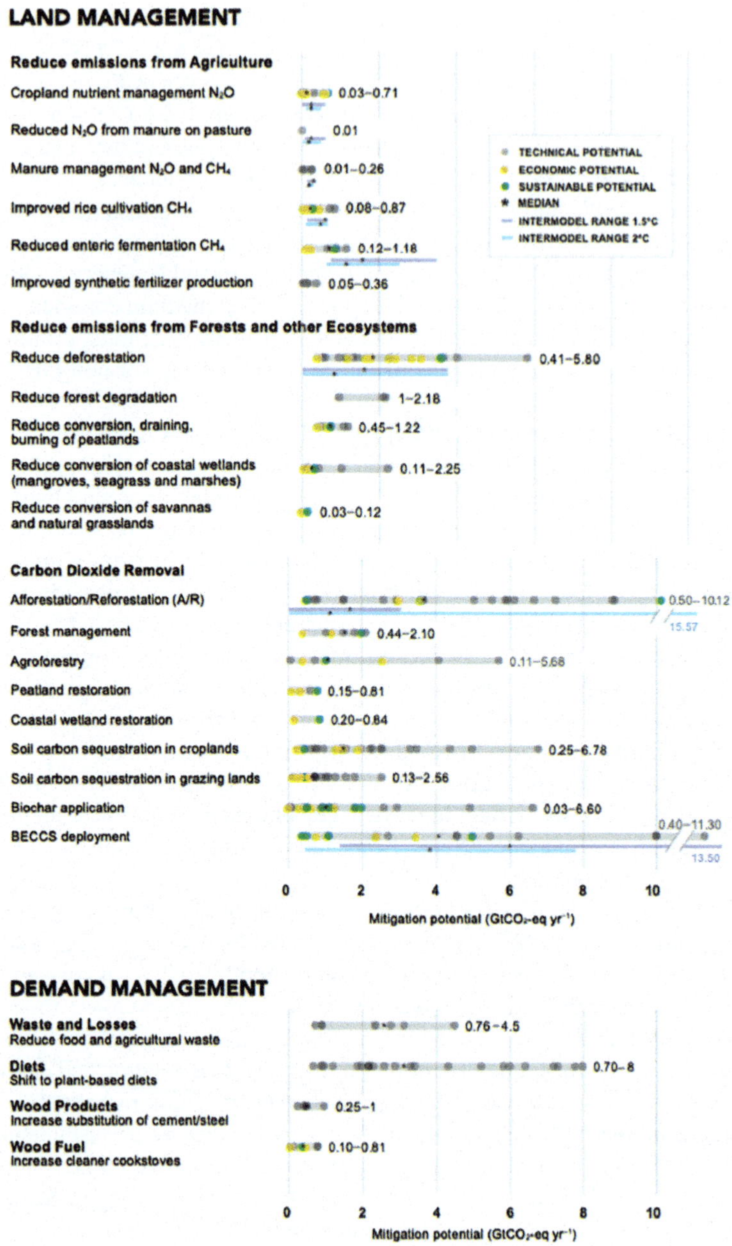

Source: (Shukla et al., 2019)

Fig. 2.8 Mitigation potential of response options in 2020–2050, measured in GtCO2-eq yr.$^{-1}$ Source: Shukla et al. (2019)

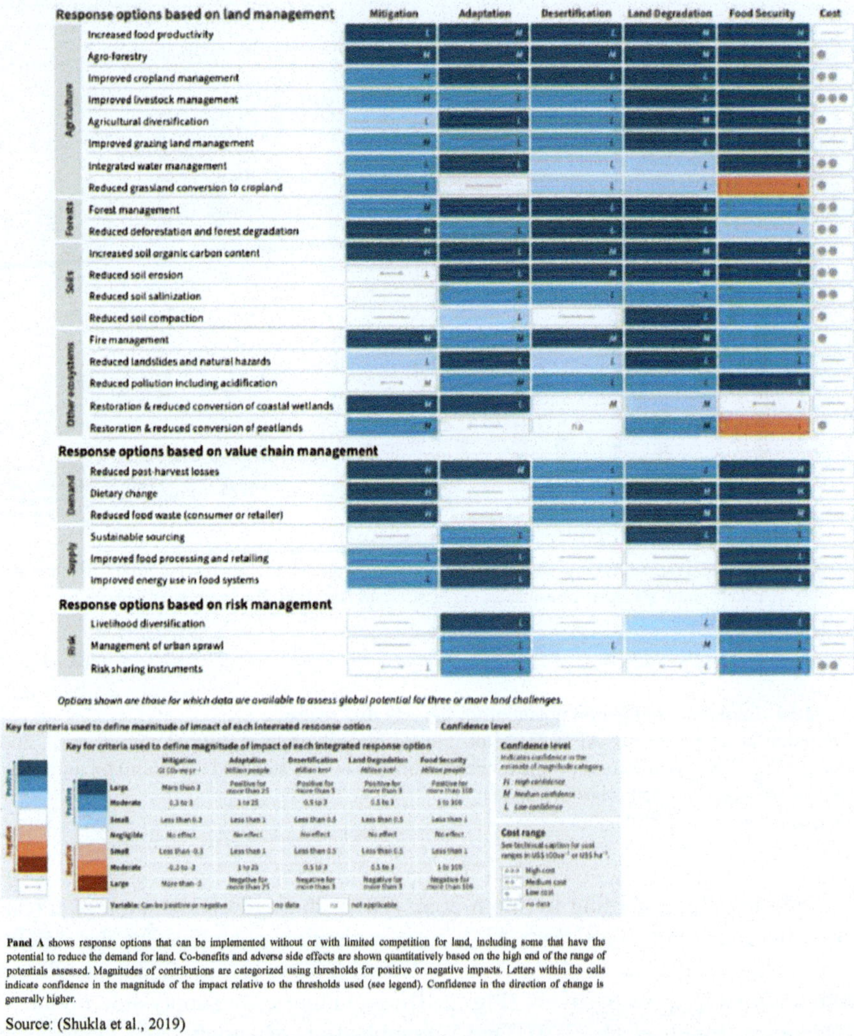

Fig. 2.9 Potential global contribution of response options to mitigation, adaptation, combating desertification and land degradation, and enhancing food security

Panel A shows response options that can be implemented without or with limited competition for land, including some that have the potential to reduce the demand for land. Co-benefits and adverse side effects are shown quantitatively based on the high end of the range of potentials assessed. Magnitudes of contributions are categorized using thresholds for positive or negative impacts. Letters within the cells indicate confidence in the magnitude of the impact relative to the thresholds used (see legend). Confidence in the direction of change is generally higher

Source: Shukla et al. (2019)

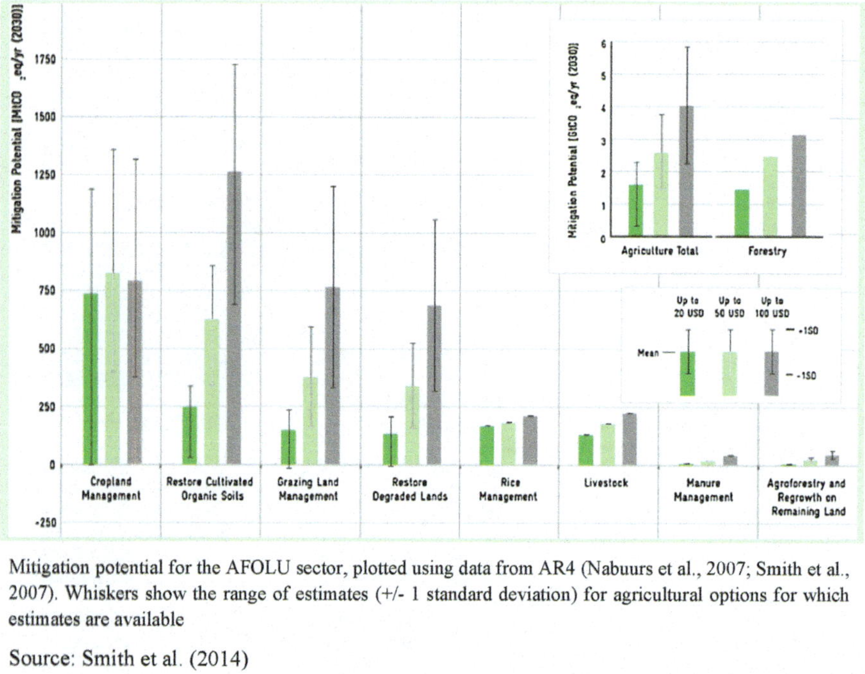

Fig. 2.10 Economic mitigation potential for the AFOLU sector
Mitigation potential for the AFOLU sector, plotted using data from AR4 (Nabuurs et al., 2007; Smith et al., 2007). Whiskers show the range of estimates (+/– 1 standard deviation) for agricultural options for which estimates are available
Source: Smith et al. (2014)

Addittionally, according to Smith et al. (2014) through combination of forestry and agriculture potentials from AR4, total mitigation potentials for the AFOLU sector are estimated to be ~3 to ~7.2 GtCO2eq/yr. in 2030 at 20 and 100 USD/tCO2eq, respectively (Fig. 2.10), including only supply-side options in agriculture (Smith et al., 2007) and a combination of supply- and demand-side options for forestry (Nabuurs et al., 2007). Figure 2.10 shows the economic mitigation potentials for the AFOLU sector until 2030. Total agriculture is distinguished from total forestry mitigation, and different agriculture mitigation options are shown for which es- timates are available (Smith et al. 2014). The mitigation options are shown at carbon prices of up to 20, 50, and 100 USD/ tCO2eq.

A comparison between different sector economic mitigation potentials has been provided in the report elaborated by the IPCC in 2007 (IPCC 2007). In which agriculture is estimated to contribute by 2030 between 2.3–6.4 GtCO2eq/yr. and forestry 1.3–4.2 GtCO2eq/yr. at <US$100/tCO2eq (IPCC 2007) (Fig. 2.11).

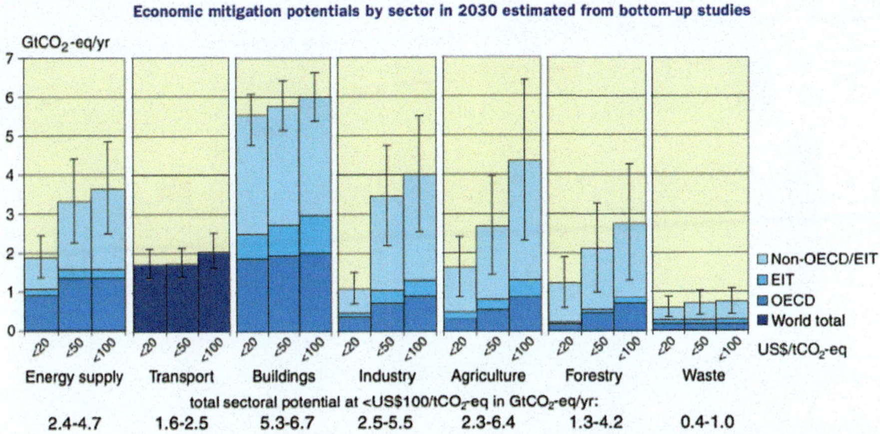

Figure SPM.10. Estimated economic mitigation potential by sector in 2030 from bottom-up studies, compared to the respective baselines assumed in the sector assessments. The potentials do not include non-technical options such as lifestyle changes. {Figure 4.2}

Notes:

a) The ranges for global economic potentials as assessed in each sector are shown by vertical lines. The ranges are based on end-use allocations of emissions, meaning that emissions of electricity use are counted towards the end-use sectors and not to the energy supply sector.

b) The estimated potentials have been constrained by the availability of studies particularly at high carbon price levels.

c) Sectors used different baselines. For industry, the SRES B2 baseline was taken, for energy supply and transport, the World Energy Outlook (WEO) 2004 baseline was used; the building sector is based on a baseline in between SRES B2 and A1B; for waste, SRES A1B driving forces were used to construct a waste-specific baseline; agriculture and forestry used baselines that mostly used B2 driving forces.

d) Only global totals for transport are shown because international aviation is included.

e) Categories excluded are: non-CO_2 emissions in buildings and transport, part of material efficiency options, heat production and co-generation in energy supply, heavy duty vehicles, shipping and high-occupancy passenger transport, most high-cost options for buildings, wastewater treatment, emission reduction from coal mines and gas pipelines, and fluorinated gases from energy supply and transport. The underestimation of the total economic potential from these emissions is of the order of 10 to 15%.

Source: (IPCC, 2007)

Fig. 2.11 Economic mitigation potential by sector in 2030 estimated from bottom-up studies Source: IPCC (2007)

Finally, a more updated study by Keramidas et al. (2018) in which trough the JRC POLES (Prospective Outlook for the Long term Energy System) model, a global sectorial simulation model for the development of energy scenarios and mitigation options for 2050, sectoral mitigation options have been analyzed (Keramidas et al. 2018).

An overview of the mitigation options adopted by the 2 °C scenario is presented in Fig. 2.12, in which were reported key mitigation options over 2015–2050 for the 2 °C scenario including ramping up the use of renewable energy sources (27%), reduction of non-CO2 emissions (20%, about a third of which are due to the decrease in fossil fuel demand in all demand sectors), improved energy efficiency (17%), electrification in final energy demand (10%) and land use (10%). Globally and by 2050, electrification exceeds one third of total final energy consumption and the share of renewables in power generation exceeds 70%.

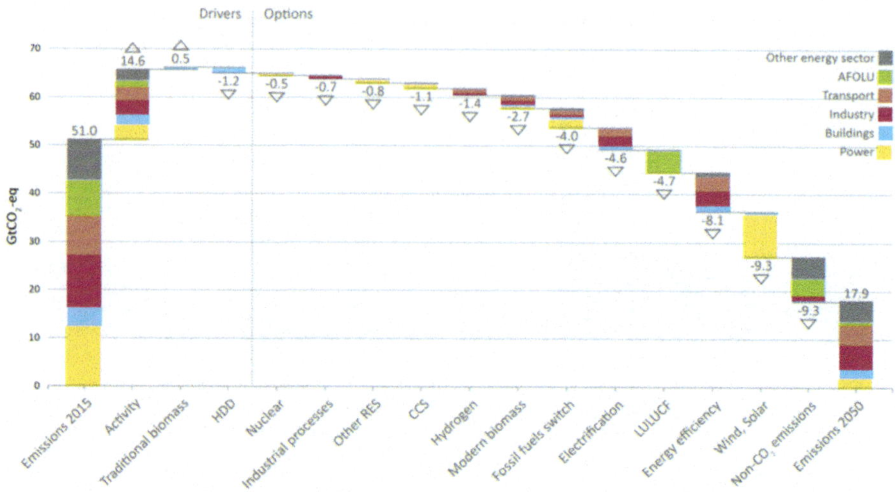

Fig. 2.12 Drivers of GHG emissions growth and mitigation, 2015–2050, 2 °C scenario, World
Source: Keramidas et al. (2018)

2.3 Mitigation Projects in the Agri-Food Sector

Considering the great potential of the agri-food sector in mitigating climate change several projects have been developed in the sector. Below we list some of these efforts:

- **REDD+**

The UN-REDD Programme is the United Nations Collaborative Programme on Reducing Emissions from Deforestation and forest Degradation (REDD+) in developing countries. The Programme was launched in 2008 and builds on the convening role and technical expertise of the Food and Agriculture Organization of the United Nations (FAO), the United Nations Development Programme (UNDP) and the United Nations Environment Programme (UNEP). Additionally, the programme supports national REDD+ readiness efforts in 65 partner countries,spanning Africa, Asia-Pacific and Latin America. The aim of REDD+ is to encourage developing countries to contribute to climate change mitigation efforts by:

- reducing greenhouse gas emissions (GHG) by slowing, halting and reversing forest loss and degradation;
- increasing removal of GHGs from the earth's atmosphere through the conservation, management and expansion of forests.

The following five REDD+ activities contribute to mitigation actions in the forest sector and have been globally agreed to:

- reducing emissions from deforestation;
- reducing emissions from forest degradation;
- conservation of forest-carbon stocks;
- enhancement of forest-carbon stocks;
- sustainable management of forests (Fig. 2.13).

- **The 4x1000 Initiative**

In addition to being sources of GHG emissions, farms also have the potential to act as "sinks" for CO2 – in other words, to remove carbon from the atmosphere and store it in soils or woody vegetation. The mitigation potential of carbon sequestration

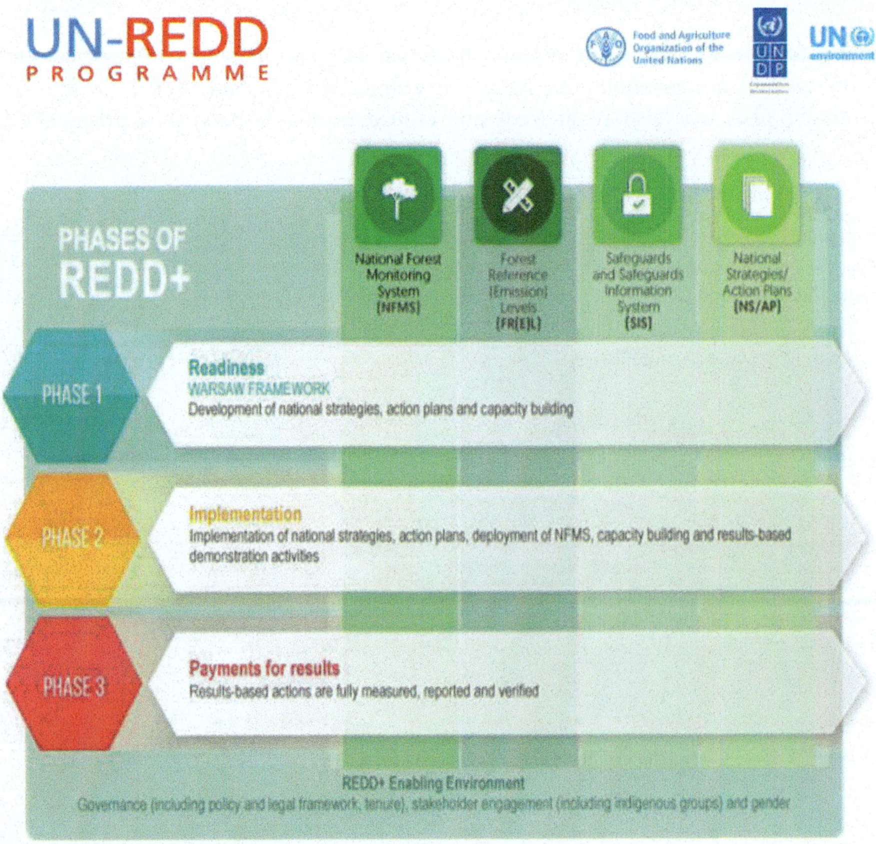

Source: (UN-REDD Programme, 2015)

Fig. 2.13 UN-REDD Programme
Source: UN-REDD Programme (2015)

in agricultural soils is large and has consequently been the focus of recent international attention such as the "4 per 1000" initiative launched at the 21st Conference of the Parties (COP 21). The 4 per 1000 initiative is an international, voluntary collaboration to increase soil organic carbon (SOC) stocks by 0.4 per cent annually, enough to halt the increase in the CO2 concentration in the atmosphere related to human activities (Minasny et al. 2017; Soussana et al. 2019).

The initiative aims to:

- Reduce deforestation;
- Encourage agro-ecological practices that increase the quantity of organic matter in soils and meet the 4 target per year (Fig. 2.14).

2.3.1 *Tree-Planting Initiatives*

Plant-for-the-Planet is an organization that aims to raise awareness amongst children and adults about the issues of climate change and global justice. The Initiative also works to plant trees and considers this to be both a practical and

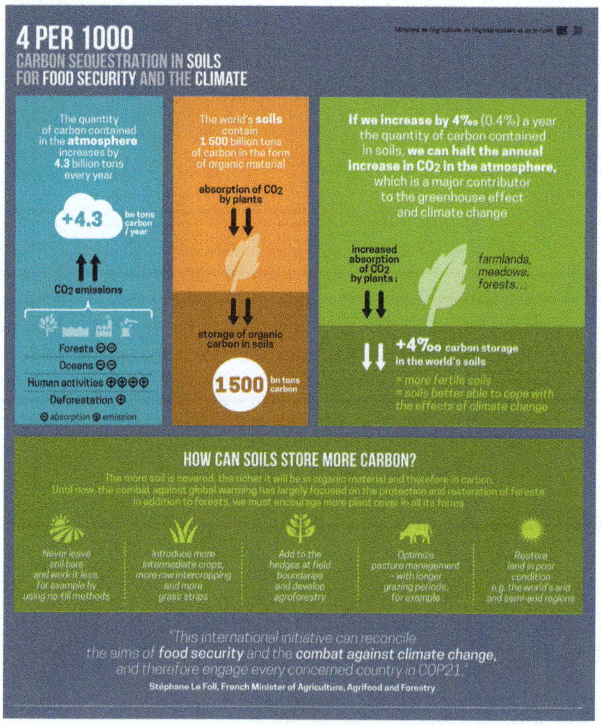

Fig. 2.14 The 4 per 1000 initiative
Source: French Ministry for Agriculture & Food (2015)

Source: https://www.plant-for-the-planet.org; https://www.bonnchallenge.org

Fig. 2.15 The plant for the planet and the Bonn challenge initiatives
Source: https://www.plant-for-the-planet.org; https://www.bonnchallenge.org

symbolic action in efforts to reduce the effect of climate change. In 2011, it reached a goal of planting a million tree.

Goal: is to plant three trillion trees currently exist globally.

The **Bonn Challenge** is a global effort to bring 150 million hectares of the world's deforested and degraded land into restoration by 2020, and 350 million hectares by 2030. It was launched in 2011 by the Government of Germany and IUCN, and later endorsed and extended by the New York Declaration on Forests at the 2014 UN Climate Summit. Underlying the Bonn Challenge is the forest landscape restoration (FLR) approach, which aims to restore ecological integrity at the same time as improving human well-being through multifunctional landscapes (Fig. 2.15).

2.3.2 Offset Projects in the Agri-Food Sector

Carbon offsetting projects are one of the economic tools available to reduce emissions by paying for metric tons of avoided CO2e emissions. Considering the mitigation potential of the agriculture and forestry sector the agri-food sector can play a pivotal role in the offsetting mechanism. Some carbon offset projects have been developed in the sector and a lot of emissions reductions cab be enabled by agricultural projects (Foucherot and Bellassen 2011).

2.3.2.1 The Fairtrade Case Study

Fairtrade International teamed up with Gold Standard and producer groups to develop the Fairtrade Climate Standard. Through projects such as reforestation or energy-efficient cookstoves, vulnerable communities can reduce emissions and become eligible for carbon credits while also strengthening themselves against the effects of climate change. Gold Standard is Fairtrade's partner in developing and implementing the Fairtrade Climate Standard and Fairtrade carbon credits. The Fairtrade Climate Standard is an add-on standard to Gold Standard certification of carbon emissions reductions and sustainable development benefits. Gold Standard is an internationally recognized organization with expertise in climate and development projects. Gold Standard supports energy, land use based and waste management projects that focus on co-benefits such as environmental benefits and local stakeholder involvement. he Fairtrade Climate Standard is open to all small-scale organized groups under Fairtrade's geographic scope. An organization developing carbon projects does not need to already be Fairtrade certified to apply.

Projects eligible for generating Fairtrade carbon credits fall into three categories:

1. **Renewable energy projects** such as solar thermal heating/electricity, solar photovoltaic, wind energy, hydropower, biogas heating/electricity
2. **Energy efficiency projects** such as improved cookstoves, water filtration/purification systems, energy saving lamps/fluorescent lamps
3. **Forestry projects** such as planting trees or replanting trees in a previously forested area

On average each project is expected to generate around 25,000 Fairtrade carbon credits per year. The length of the projects depends on the type of project – energy projects may run for 7–10 years; forestry projects may run for 30–40 years (Fig. 2.16).

2.3.2.2 The Rainforest Alliance Case Study

Rainforest Alliance is auditing forestry and agroforestry projects in which landowners, forest communities, or farmers plant trees on degraded land (reforestation), plant trees where forests are long gone (afforestation), and/or protect standing forests from being cut down. As an ISO 14065 accredited verification body, the Rainforest Alliance conducts independent audits of forestry projects designed to sequester carbon dioxide and reduce GHG emissions. This means that they verify the accuracy of a project's emission reductions and carbon credits claims. As of May 2014, they have conducted validations (of project design) or verifications (of actual emissions

Fig. 2.16 Fairtrade Carbon Credits
Source: https://www.fairtrade.net/product/carbon-credits

Fig. 2.17 Rainforest Alliance Logo
Source: https://www.rainforest-alliance.org/articles/understanding-tools-to-fight-climate-change

Source: https://www.rainforest-alliance.org/
articles/understanding-tools-to-fight-climate-change

sequestered) for more than 50 different carbon projects, spanning 23 countries and representing more than 4 million hectares of land under conservation/ improved forest management. The Rainforest Alliance has also validated that these projects have the potential to result in 170,013,495 tCO2e from avoided emissions or sequestration. The Rainforest Alliance also provides technical support to carbon standards, project developers and businesses and is actively involved in facilitating the development of REDD+ and zero deforestation projects in countries like Mexico, Ecuador, and Peru. Their aim is to foster sustainable forest management and land-use practices as building blocks for creating high-value forest carbon credits. Much of this work focuses on building the capacities of communities to implement REDD, through training and technical assistance, and working to shape REDD+ and climate policies in such a way that ultimately helps countries better prepare for global warming (Fig. 2.17).

References

Altieri MA, Nicholls CI (2017) The adaptation and mitigation potential of traditional agriculture in a changing climate. Clim Chang. https://doi.org/10.1007/s10584-013-0909-y

European Commission (2015) EU agriculture and climate change - factsheet. Eur. Comm

Foucherot C, Bellassen V (2011) Carbon offset projects in the agricultural sector. Carbon N. Y

French Ministry for Agriculture & Food (2015) Join the 4 per 1000 initiative: soils for food security and climate change, Paris Climate Change Conference

Golub A, Hertel T, Lee HL, Rose S, Sohngen B (2009) The opportunity cost of land use and the global potential for greenhouse gas mitigation in agriculture and forestry. Resour Energy Econ. https://doi.org/10.1016/j.reseneeco.2009.04.007

Golub AA, Henderson BB, Hertel TW, Gerber PJ, Rose SK, Sohngen B (2013) Global climate policy impacts on livestock, land use, livelihoods, and food security. Proc Natl Acad Sci U S A. https://doi.org/10.1073/pnas.1108772109

IPCC (2007) Climate change 2007 synthesis report, Intergovernmental Panel on Climate Change [Core Writing Team IPCC]. https://doi.org/10.1256/004316502320517344

IPCC (2018) Summary for policymakers, global warming of 1.5 °C. An IPCC special report on the impacts of global warming of 1.5 °C above pre-industrial levels and related global greenhouse gas emission pathways, in the context of strengthening the global response to the threat of climate change,

Keramidas K, Tchung-Ming S, Diaz-Vazquez AR, Weitzel M, Vandyck T, Després J, Schmitz A, Rey Los Santos L, Wójtowicz K, Schade B, Saveyn B, Soria-Ramirez A, Forschungsstelle EKG (2018) Global energy and climate outlook 2018: sectoral mitigation options towards a low-emissions economy – CORE. Publications Office of the European Union. https://doi.org/10.2760/67475

Mayhew AL (2016) The state of food and agriculture. Climate change, agriculture and food security, Food and Agriculture Organization of the United Nations (FAO). https://doi.org/10.1093/nq/s8-IV.94.301-a

Minasny B, Malone BP, McBratney AB, Angers DA, Arrouays D, Chambers A, Chaplot V, Chen ZS, Cheng K, Das BS, Field DJ, Gimona A, Hedley CB, Hong SY, Mandal B, Marchant BP, Martin M, McConkey BG, Mulder VL, O'Rourke S, Richer-de-Forges AC, Odeh I, Padarian J, Paustian K, Pan G, Poggio L, Savin I, Stolbovoy V, Stockmann U, Sulaeman Y, Tsui CC, Vågen TG, van Wesemael B, Winowiecki L (2017) Soil carbon 4 per mille. Geoderma. https://doi.org/10.1016/j.geoderma.2017.01.002

Nabuurs GJ, Masera O, Andrasko K, Benitez-Ponce P, Boer R, Dutschke M, Elsiddig E, Ford-Robertson J, Frumhoff P, Karjalainen T, Krankina O, Kurz WA, Matsumoto M, Oyhantcabal W, Ravindranath NH, Sanchez MJS, Zhang X (2007) Forestry. In: Metz B, Davidson OR, Bosch PR, Dave R, Meyer LA (Eds.): Climate Change 2007. Mitigation. Contribution of Working Group III to the Fourth Assessment Report of the Intergovernmental Panel on Climate Change. Cambridge University Press, Cambridge, United Kingdom and New York, USA

Nauclér T, Enkvist P (2009) Pathways to a low-carbon economy: version 2 of the global greenhouse gas abatement cost curve. McKinsey & Company. https://doi.org/10.1016/j.enpol.2010.01.047

OECD (2019) Enhancing climate change mitigation through agriculture. https://doi.org/10.1787/e9a79226-en

Poore J, Nemecek T (2018) Reducing food's environmental impacts through producers and consumers. Science (80-.). 360, 987–992. https://doi.org/10.1126/science.aaq0216

Richards M, Arslan A, Cavatassi R, Rosenstock T (2019) Climate change mitigation potential of agricultural practices supported by IFAD investments: an ex ante analysis

Rose SK, Ahammad H, Eickhout B, Fisher B, Kurosawa A, Rao S, Riahi K, van Vuuren DP (2012) Land-based mitigation in climate stabilization. Energy Econ. https://doi.org/10.1016/j.eneco.2011.06.004

Shukla PR, Skea J, Buendia EC, Masson-Delmotte V, Pörtner H-O, Roberts DC, Zhai P, Slade R, Connors S, van Diemen R, Ferrat M, Haughey E, Luz S, Neogi S, Pathak M, Peold J, Pereira JP, Vyas P, Huntley E, Kissick K, Belkacemi M, Malley J (2019) Climate change and land: an IPCC special report on climate change, desertification, land degradation, sustainable land management, food security, and greenhouse gas fluxes in terrestrial ecosystems. IPCC. https://doi.org/10.4337/9781784710644

Smith P (2012) Agricultural greenhouse gas mitigation potential globally, in Europe and in the UK: what have we learnt in the last 20 years? Glob Chang Biol. https://doi.org/10.1111/j.1365-2486.2011.02517.x

Smith P (2016) Soil carbon sequestration and biochar as negative emission technologies. Glob Chang Biol. https://doi.org/10.1111/gcb.13178

Smith P, Martino D, Cai Z, Gwary D, Janzen H, Kumar P, McCarl B, Ogle S, O'Mara F, Rice C, Scholes B, Sirotenko O, Howden M, McAllister T, Pan G, Romanenkov V, Schneider U, Towprayoon S (2007) Policy and technological constraints to implementation of greenhouse gas mitigation options in agriculture. Agric Ecosyst Environ. https://doi.org/10.1016/j.agee.2006.06.006

Smith P, Martino D, Cai Z, Gwary D, Janzen H, Kumar P, McCarl B, Ogle S, O'Mara F, Rice C, Scholes B, Sirotenko O, Howden M, McAllister T, Pan G, Romanenkov V, Schneider U, Towprayoon S, Wattenbach M, Smith J (2008) Greenhouse gas mitigation in agriculture. Philos Trans R Soc B Biol Sci. https://doi.org/10.1098/rstb.2007.2184

Smith P, Bustamante M, Ahammad H, Clark H, Dong H, Elsiddig EA, Haberl H, Harper R, House J, Jafari M, Masera O, Mbow C, Ravindranath NH, Rice CW, Abad CR, Romanovskaya A, Sperling F, Tubiello F (2014) Agriculture, forestry and other land use (AFOLU), in: climate change 2014: mitigation of climate change. Contribution of Working Group III to the Fifth Assessment Report of the Intergovernmental Panel on Climate Change. Cambridge University Press, Cambridge, United Kingdom and New York, NY, USA, pp 811–922

Soussana JF, Lutfalla S, Ehrhardt F, Rosenstock T, Lamanna C, Havlík P, Richards M, Wollenberg E (Lini), Chotte JL, Torquebiau E, Ciais P, Smith P, Lal R (2019) Matching policy and science: rationale for the '4 per 1000 - soils for food security and climate' initiative. Soil Tillage Res. https://doi.org/10.1016/j.still.2017.12.002

UN-REDD Programme (2015) UN-REDD Programme strategic framework 2016–20 20, 52

Vermeulen SJ, Campbell BM, Ingram JSI (2012) Climate change and food systems. Annu Rev Environ Resour. https://doi.org/10.1146/annurev-environ-020411-130608

Wieben E (2019) Priorities related to food value chains and the Agri-food sector in the nationally determined contributions (NDCs). FAO

Chapter 3
Putting a Farm into an Emissions Test Lab

The third chapter explores the carbon footprint methodology. Particularly, it presents both methodologies and standards to calculate the Organizational carbon footprint (OCF) and the Product Carbon Footprint (PCF). Moreover, the chapter presents the main methodological issues related to carbon accounting in the agri-food sector.

3.1 Life Cycle Assessments (LCA) and Carbon Footprints

Life Cycle Assessments (LCA) and Carbon Footprints are research methods to gain insight in the full life cycle of a product. It quantifies all inputs and outputs. This means all phases of a product are taken into account and analyzed, from raw materials, packaging, transport, retail and consumer towards disposal or recycling. This means a product will be studied from cradle-to-grave. An LCA shows where the environmental impact in the supply chain takes place. Hotspots are the locations in the supply chain with the highest environmental impact. Disclosure of these hotspots is an opportunity to improve a certain process or product. Eventually, this will lead towards savings in for instance raw materials, energy use and cost and a more sustainable product. In some cases, the environmental advantages can also be used as selling point (Fig. 3.1).

3.2 Carbon Footprint (CF)

A Carbon Footprint (CF) is a means for measuring, managing and communicating greenhouse gas (GHG) emissions related to organizations, goods and services. A carbon footprint is based on a life cycle assessment (LCA) but focuses on a single issue which is global warming.

A. Acampora et al., *Carbon Neutrality in the Agri-food Sector*,
https://doi.org/10.1007/978-3-030-88048-4_3

Fig. 3.1 Life Cycle Assessment and Footprint
Source: Barilla Center for Food and Nutrition (2016)

There are two types of carbon footprint:

- **Organisational carbon footprint**
- **Product/service carbon footprint**

3.2.1 Organizational Carbon Footprint (OCF)

An organizational carbon footprint measures GHG emissions from all activities across a company. This includes energy used in buildings and industrial processes, company owned vehicles and may measure indirect emissions associated with activities outside an organization's own operations - the value chain. Value chain analysis looks at every step a business goes through, from raw materials to the eventual end-user.

The main standard to assess the OCF are:

- **ISO 14064-1:**ISO 14064 standard was published in 2006 and is part of the ISO series of International Standards for environmental management. This international standard provides guidance on the principles and requirements for reporting GHG emissions. It provides additional guidance on verification, required levels of data validation and external reporting frameworks, to ensure consistent external communication.
- **GHG Protocol Corporate Standard:** The GHG protocol is an emissions accounting tool used by many businesses and organizations worldwide. The

standard was developed to address the need for a consistent approach in corporate carbon accounting and reporting. The Corporate Standard categorizes: GHG emissions into 3 scopes. Scope 1 (Direct emissions that result from activities within your organization's control), Scope 2 (Indirect emissions from any electricity, heat or stream you purchase and use) and Scope 3 (other indirect emissions from sources outside your direct control).
- **GHG Protocol Value Chain Standard:** This standard accompanies the GHG Protocol corporate Standard. It provides additional guidance to companies wanting to assess their entire value chain Scope 3 emissions. Emissions are grouped into 15 categories of scope 3 activities, both upstream and downstream. Particularly useful for companies that want to report detailed information on their Scope 3 value chain emissions externally. For example, companies reporting to CDP are required to provide extensive value chain emissions assessments.

3.2.2 Product Carbon Footprint (PCF)

A product carbon footprint measures the GHG emissions over the life of a product or service. This involves calculating emissions from the extraction of raw materials and manufacturing, through to emissions associated with the use and disposal of a particular product.

The main standard to assess the PCF are:

- **ISO 14067:** Purpose of the standard is to increase the transparency in reporting GHG emissions associated with the entire lifecycle of different products and services. ISO 14067 is aligned with previous ISO and PAS 2050 product carbon footprinting standards to increase the comparability of product carbon footprints internationally. The standard includes detailed guidance on requirements for public reporting and external communication, as well as additional guidance on verification and assurance of product carbon footprints.
- **GHG Protocol Product Lifecycle Accounting and reporting standard:** Launched in October 2011 after a 3-year multi stakeholder development process. This standard builds upon existing ISO environmental guidance and aims to provide a general framework for accounting and reporting product lifecycle GHG emissions. Public reporting is required to claim conformance to the standard.
- **PAS 2050:** PAS 2050 was developed by the British Standards Institute in response to a desire for a consistent method for assessing the life cycle GHG emissions associated with products or services. The standard is widely recognized, internationally applied and provides a consistent method for assessing product life cycle GHG emissions. The standard can be used on a wide range of product and service types including, goods and services, business-to-consumer, manufacturers, retailers and traders. PAS 2050 does not set requirements for product carbon footprint external communication.

Table 3.1 presents a comparative analysis of both the different Carbon Footprint standards, LCA standards and Carbon neutrality standard.

3.3 The GHG Protocol Initiative

GHG Protocol is private international standard that establishes comprehensive global standardized frameworks to measure and manage greenhouse gas (GHG) emissions from private and public sector operations, value chains and mitigation actions. Building on a 20-year partnership between World Resources Institute (WRI) and the World Business Council for Sustainable Development (WBCSD), GHG Protocol works with governments, industry associations, NGOs, businesses and other organizations.

Companies and Organizations
In 2016, 92% of Fortune 500 companies responding to the CDP used GHG Protocol directly or indirectly through a program based on GHG Protocol.

Countries and Cities
Through their commitment to the Compact of Mayors, hundreds of cities across the globe have committed to using the GHG Protocol for Cities.

They also work with partners in key countries to develop national GHG emissions programs based on the GHG Protocol (Fig. 3.2).

3.3.1 The GHG Protocol for Guidance on Carbon Removals and Land Use

The Greenhouse Gas Protocol is launching a process to develop new standards or guidance on how companies should account for the following activities in their greenhouse gas inventories:

- Carbon removals and sequestration
- Land use
- Land use change
- Bioenergy
- Related topics

The project will develop internationally accepted standards or guidance on corporate GHG accounting on the above topics. New standards or guidance are expected to be used by companies to:

- Inform mitigation strategies by understanding the GHG emissions/removals impacts of land use, land use change, bioenergy and carbon removal activities

Table 3.1 Comparison of Carbon related standards

Name	Title	Owner	Carbon footprint	LCA	Removals; GHG sinks	Label/ reporting guidelines	Scope
LCA Methodology related standard							
ISO 14040	Environmental management -- Life cycle assessment -- Principles and framework	ISO	X	V	X	X	
ISO 14044:2006	Environmental management -- Life cycle assessment -- Requirements and guidelines	ISO	X	V	X	X	
ISO 14072	Environmental management -- Life cycle assessment -- Requirements and guidelines for organizational life cycle assessment	ISO	V	V	X	X	
Communication							
ISO 14064-3	Greenhouse gases Specification with guidance for the verification and validation of greenhouse gas statements	ISO	V	V	V	V	
Product Carbon Footprint							
ISO 14067	Greenhouse gases -- Carbon footprint of products -- Requirements and guidelines for quantification	ISO	V	V	X	X	Cradle-to-grave; Cradle-to-gate; Gate-to-gate and partial life cycle
PAS 2050	Specification for the assessment of the life cycle greenhouse gas emissions of goods and services	BSI	V	V	V	V	Cradle-to-gate and cradle-to-grave
GHG Protocol	Product Life Cycle Accounting and Reporting Standard	WRI/ WBCSD	V	V	V	V	cradle-to-gate and cradle-to-grave

(continued)

Table 3.1 (continued)

Name	Title	Owner	Carbon footprint	LCA	Removals; GHG sinks	Label/ reporting guidelines	Scope
Organizational Carbon Footprint							
ISO 14064-1	Greenhouse gases Specification with guidance at the organization level for quantification and reporting of greenhouse gas emissions and removals	ISO	V	V	V	V	Direct, energy indirect, Other indirect
GHG Protocol	A Corporate Accounting and Reporting Standard	WRI/WBCSD	V	V	X	V	Scope 1; 2; 3
GHG Protocol	Corporate Value Chain (Scope 3) Accounting and Reporting Standard	WRI/WBCSD	V	V	X	V	Scope 3
Bilan Carbone®	Emissions factors and their calculation to allow GHG reporting under the Bilan CarboneTM Method	Agence de l'Env. et de la Maîtrise de l'Energie (ADEME)	V	X	X	V	Direct, energy indirect, Other indirect
Project Carbon Footprint							
ISO 14064-2	Greenhouse gases Specification with guidance at the project level for quantification, monitoring and reporting of greenhouse gas emission reductions or removal enhancements	ISO	X	V	V	V	Offsetting projects
GHG Protocol	The GHG Protocol for Project Accounting	WRI/WBCSD	X	V	V	V	Offsetting projects
Carbon Neutrality							
PAS 2060	Specification for the demonstration of carbon neutrality	BSI	V	V	Y	V	Scope 1; 2; 3

Source: Authors' own elaboration

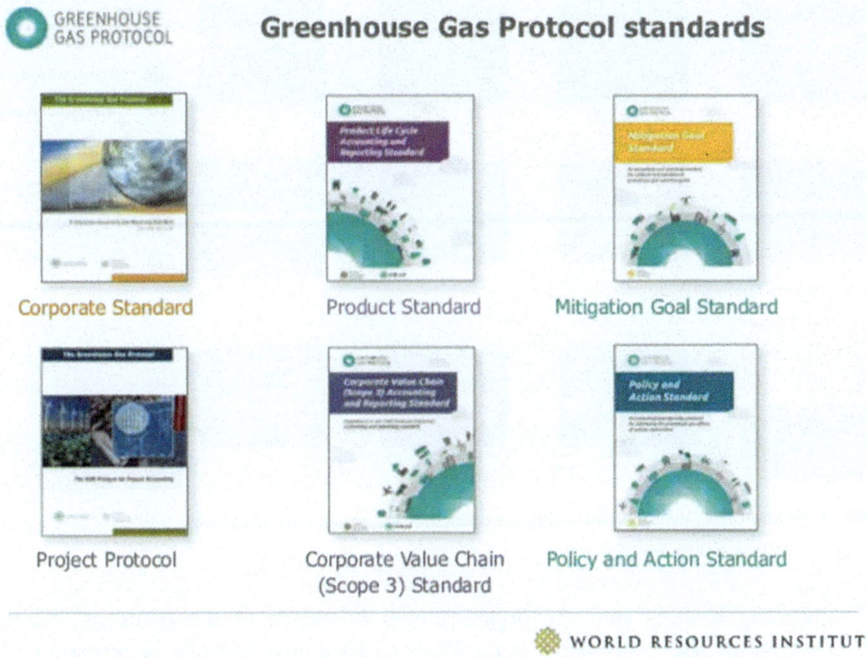

Fig. 3.2 Different GHG protocol standards
Source: http://ghgprotocol.org

- Set targets and track performance by including the above activities in GHG targets
- Report GHG inventories including GHG emissions and carbon removals and report progress toward GHG mitigation goals (Fig. 3.3)

3.3.2 Scope of the GHG Protocol Corporate Standard

The GHG Protocol Corporate Standard classifies a company's GHG emissions into three 'scopes':

- **Scope 1- Direct GHG emissions –** these occur from sources that are owned or controlled by the company, for example emissions from combustion in owned or controlled boilers, furnaces, vehicles, etc. or emissions from chemical production in owned or controlled process equipment.
- **Scope 2- Electricity and heat indirect GHG emissions –** this accounts for GHG emissions from the generation of purchased electricity and heat consumed by the company. Purchased electricity is defined as electricity that is purchased or

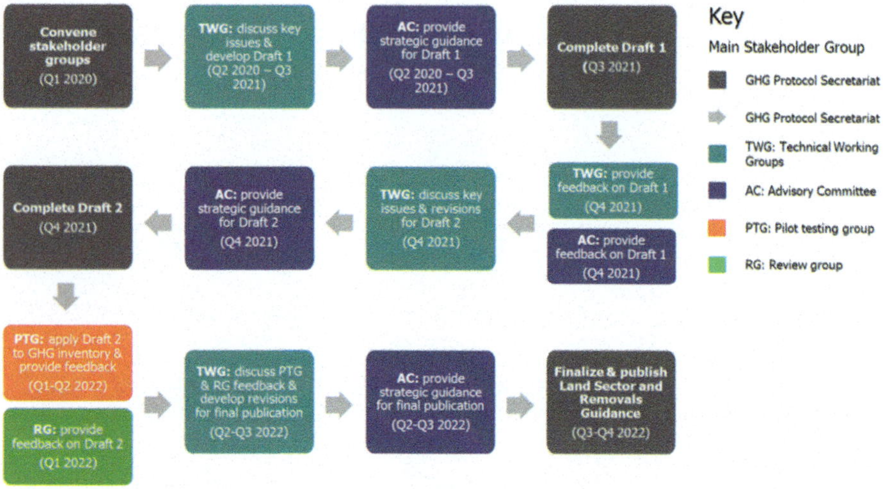

Fig. 3.3 The GHG protocol for guidance on carbon removals and land use
Source: http://ghgprotocol.org

otherwise brought into the organizational boundary of the company. Scope 2 emissions physically occur at the facility where the electricity is generated.

- **Scope 3 - Other indirect GHG emissions** – this is a reporting category that allows for the treatment of all other indirect emissions. Scope 3 emissions are a consequence of the activities of the company but occur from sources not owned or controlled by the company. Some examples of Scope 3 activities are the extraction and production of purchased materials, the transportation of purchased fuels and the use of sold goods and services (Fig. 3.4).

Apply Scope 1-2 in the agri-food sector in relatively easy. However, apply scope 3 can presents some issue that are mostly related to the absence of primary data geographically and temporally focused and difficulties in accounting GHG emissions and removals for:

- Yields
- Fertilizers and Pesticides
- Agricultural measures

3.4 Carbon Footprint in Agri-Food Sector: Methodological Issues

None of these standards are specifically designed for the agri-food sector. According to Notarnicola et al. (2016, 2017) the current LCA method is **incomplete** and **does not comprehensively assess some aspects that are critical for long-term**

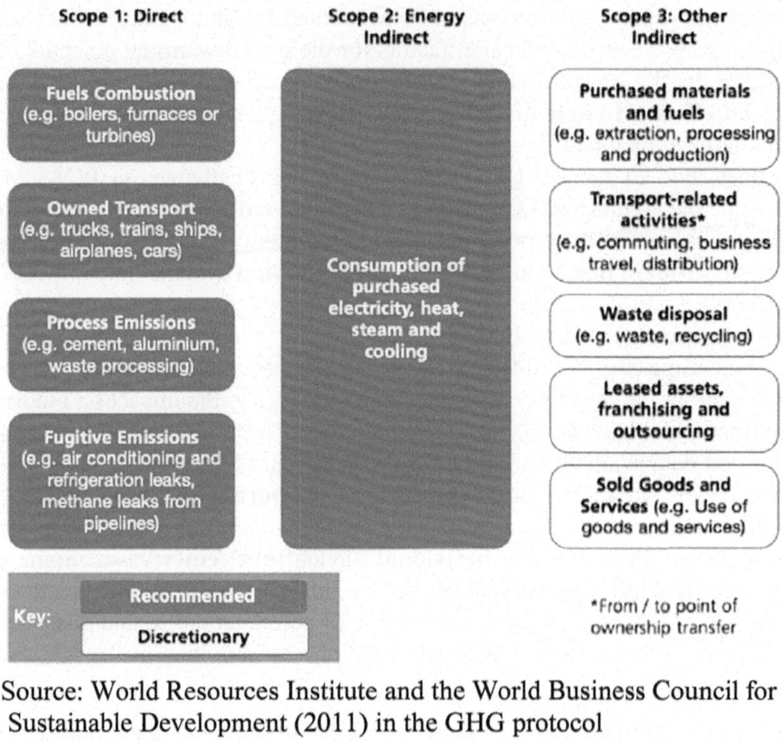

Fig. 3.4 Scope of the Organizational Carbon footprint according to the GHG protocol
Source: World Resources Institute and the World Business Council for Sustainable Development (2011) in the GHG protocol

sustainable food production (e.g. decreased soil quality and fertility, increased erosion, reduced ecosystem services due to intensification, biodiversity loss). In the following paragraphs are reported the main methodological issues underlined by Notarnicola et al. (2016, 2017).

The **landscape pattern** is fundamental to understand the level of **threat to biodiversity posed by an agricultural system**. This is information at landscape level which is completely missing in LCA. LCA studies usually focus on the field level and therefore, by not acknowledging such emergent aspects, the conclusions from an LCA study might support less preferred policies and actions from a sustainability perspective. LCA-based methods that combine both the field and landscape perspectives are needed in order to capture environmental impacts at these different scales.

Several **agricultural measures for mitigating impacts** (such as having field margins acting as buffer zones) **are not captured by current land use inventories, whereas reporting their presence may be regarded as a "credit" in terms of reduction of environmental impacts**. Also, agricultural practices such as composting and reduced tillage can have a **carbon sequestration** effect which can vary over the years and hence has to be accounted for and averaged over a period consistent with the characterization factors for the global warming potential (GWP) considered in the specific study. Furthermore, such practices can also have a **beneficial effect on biodiversity**, which is **seldom accounted for in LCA**s since no standardized methodology is available.

Besides, two of the most incomplete modelling challenges in LCAs of food systems are the significant **inconsistencies between emission inventory modelling and impact assessment of pesticides**, and **assessment of land use change associated with off-farm inputs to agricultural production systems** (Notarnicola et al. 2016, 2017).

Food systems are inherently more variable in the inventory data (e.g. in the same area, on the same crop, two different active ingredients could be applied for the same purpose) and in the reliability of impact assessment (e.g. the impact on biodiversity due to a particular land use may change dramatically from one ecoregion to another). Yet, current data available in food LCA databases and life cycle impact assessment (LCIA) models, are mostly **non-spatially and temporally resolved.**

In particular, for agricultural systems, there is potential for considerable variability in inventory data **between individual agricultural enterprises**. Some of the aspects that underlie this variability include different management practices, soil types and climates, seasonality, the life cycle of perennial crops, and distances (and related transportation modes) between locations of activities in the life cycle of product systems.

- Regarding **management practices**, farmers may utilize different practices based on their own preferences and expertise. Sets of practices may be grouped together and labelled as "organic", "biodynamic", "integrated", "heated greenhouse" production. However, in reality there is usually a **continuous spectrum of practices both within and between these categories**. Beyond human-controlled variability, differences in management practices and in yields may be related to **soil types and climates** e.g. some soils require regular application of lime to raise the soil's pH, and dryer climates require use of irrigation systems.
- Some sources of variability are related to the **timescale adopted for the study**. Within a single year, **seasonality** may contribute to differences in LCA results for food products. Between years the environmental impacts of a single crop may vary due to differences in yields related **to variable weather conditions**. Furthermore, over a period of several years, **perennial crops exhibit a cycle of increasing and then decreasing yields and this is often not accounted for in LCA studies**

Finally, variability may be related to the different **transport distances (and modes)** between the locations of agricultural production in relation to production of inputs used in agricultural production (such as fertilizers and compost), and subsequent processing, retailing and consumption activities.

Moreover, impacts on **soil fertility, soil structure, soil hydrology balance, and soil biodiversity** are currently not included in the majority of food LCA studies despite being essential elements for ensuring the conservation of the natural capital as well as long- term security of food supply.

Other methodological issue related to emission calculation in the agri-food sector are related to:

- **Functional Unit**: Performing LCAs for the same product system with different functional units (FU), each based on a single different function of the product, has often been shown to generate completely different or even contrasting LCA results. The selection of the FU is thus critical as it should best represent the scope of the assessment.
- **Pesticides**: Different models used at inventory analysis provide different environmental results for pesticides. The destination of a pesticide is not only dependent on the plant but also on site and time issues such as the type of soil, the weather conditions, the location of the water table, and farming practices.
- **Fertilizers**: In relation to emissions from use of fertilizers a consensus is still missing on a globally applicable model for calculating soil and water emissions (i.e. leaching, erosion and run- off) which are more dependent on soil conditions (e.g. pH, clay content, slope, etc.).
- **Databases**: One of the key issues in the agri-food sector is the lack of reliable and up-to-date inventory data on food products and processes for realizing not only accurate LCA studies but also for hotspot analysis, communication, and labelling. Several LCI databases have been developed but most of them are characterized by a lack of transparency, and they are often incomplete because they take into account only a few input-output flows; this lack of information concerning food-product impacts can lead to ambiguous interpretations and conclusions; furthermore, these databases are often outdated and not regionalized. Finally, the databases are often inconsistent with each other, because of their different approaches and assumptions. Comprehensive, clear, well-documented, and consistent data are needed to in- crease the accuracy and comparability of LCAs in the food sector.
- **End-of-life**: In LCAs the end-of-life phase is often omitted, thus excluding an important means of making more complete the evaluation of environmental impacts because of the omission of important results and failure to consider the possibilities of making a product system more recyclable or disposable.
- **Industrial phase**: data concerning the industrial phases of food production are not always accessible. Industrial companies are not always willing to reveal information about their industrial processes. Assumptions and estimates may therefore have to be made which do not reflect the reality of the product system under analysis.

Moreover, data regarding **pesticides and fertilizers production** are not always available. This leads either to the use of estimates for the assessment or to the complete exclusion of such a phase from the study. Similar issues arise for emissions from the **use of machinery** where fuel consumption is dependent not only on hours of work but also on aspects such as tractor power, type of operation and soil conditions.

Furthermore, although **water and land use in agriculture** could have major environmental consequences, most LCA studies represent these impacts as mere flows expressed in m2 or m3 and do not assess the potential environmental damage arising from these uses. The assessment methodologies of the impacts of land and water use in LCAs are by no means standardized and are evolving constantly. Water and land use assessment are also site-specific and hence there is a need for regionalized datasets.

Also, potential for significant variability to arise at other stages in the life cycle of food products. In particular, this variability may be related to **storage time, packaging and food preparation** (including related wastage).

References

Barilla Center for Food and Nutrition (2016) DoUBLE PYRAMID 2016 Eat better Eat less Food for all

Notarnicola B, Salomone R, Petti L, Renzulli PA, Roma R, Cerutti AK (2016) Life cycle assessment in the Agri-food sector: case studies, methodological issues, and best practices. Int J Life Cycle Assess. https://doi.org/10.1007/s11367-015-0977-5

Notarnicola B, Sala S, Anton A, McLaren SJ, Saouter E, Sonesson U (2017) The role of life cycle assessment in supporting sustainable Agri-food systems: a review of the challenges. J Clean Prod 140:399–409. https://doi.org/10.1016/j.jclepro.2016.06.071

Chapter 4
Governance Issues in Carbon Emissions: Can a Food Emissions Gate Be Avoided?

The fourth chapter investigates the problem of greenwashing and the importance of sustainability reporting and carbon disclosure. In this sense, it explores different carbon disclosure initiatives underling the importance of good governance for triggering possible drivers of carbon neutrality investment and 'making markets work'.

4.1 'Emissions Gate' and Increasing Concerns Over Greenwashing

However, the major challenge that companies, marketers and governments have to deal with is the consumers' "greenwashing" skepticism. Greenwashing refers to: "*intentionally misleading or deceiving consumers with false claims about a firm's environmental practices and impact*" (Wang 2017). When consumers feel that a company claim is "greenwashed", they lose trust in that brand compromising brad equity. It is, therefore, crucial that companies' efforts are seen by consumers as a genuine commitment to sustainability and climate change mitigation.

According to Horiuchi et al. (2009) there are different types of greenwashing:

- **Effectiveness of Communications:** This category includes companies that have made substantial efforts to improve the environmental performance of their products and processes but are unable to communicate these efforts effectively.
- **Unsubstantiated Greenwash:** At first glance, these companies seem to be doing commendable work and providing data to back up their claim. However, a deeper dive shows that the company does not deserve as much credit as it seems. It may be lobbying against the very environmental policies it claims to uphold, or it may be putting more resources into its communications than its actual initiatives.
- **Greenwash Noise:** In cases where a company says, "we're green," but does not have much to back up this claim, these messages are not compelling to

A. Acampora et al., *Carbon Neutrality in the Agri-food Sector*,
https://doi.org/10.1007/978-3-030-88048-4_4

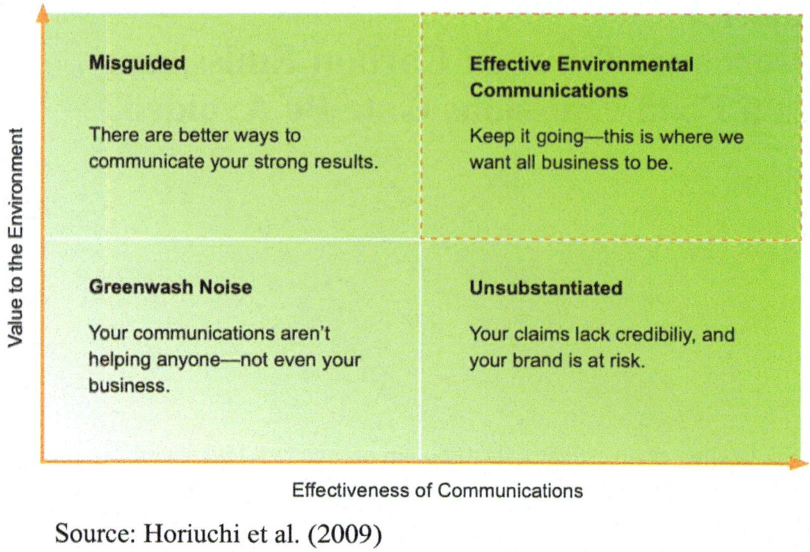

Fig. 4.1 Types of greenwash
Source: Horiuchi et al. (2009)

consumers. Much work needs to be done to move these companies to the top right quadrant, but it is feasible. By assessing the company's impacts throughout the value chain, developing and implementing an environmental strategy, and then communicating these efforts accurately, these companies can create a path to the "Effective Environmental Communications" quadrant.
- **Effective Environmental Communications:** This is the goal we have for all companies. These businesses are improving the environmental and social performance of their products and aligning these efforts throughout various functions within the company. They are able to communicate their efforts so that consumers clearly understand the impacts and other businesses look to these companies for leadership.

These types of greenwashing can be visualized trough the matrix in Fig. 4.1.

These barriers are even more critical when considering consumers perception of carbon neutral concept. Especially for what concern the practice of carbon offsetting, this has often been seen by consumers as a greenwashing practice and generally perceived with skepticism (Bunning et al. 2013). This criticism regarding carbon offset has led to the diffusion of the term "carbon indulgence". Offset have been seen as a way for polluting companies to wash their sins by paying an indulgence, the offset credit, (as in indulgences of the Middle Ages) and continue to pollute (Nerlich and Koteyko 2009). Offset practices are also been criticized as they have been seen as a way for companies to pay to pollute rather than adjusting their production to more eco-efficient solutions and to reduce their emissions (Hyams and Fawcett 2013).

Additionally, Birkenberg and Birner (2018) pointed out that further reasons for this negative image are:

- a confusing set of related terms, such as zero-carbon, carbon-neutral, carbon- free and climate-neutral;
- lack of transparency on how carbon reductions were achieved and where the carbon credits came from;
- a lack of comparability due to missing standardized methodologies and differences in functional units.

Another problem is related to the fact that many green groups continue to be sceptical about the true impact of carbon offsetting, with questions around issues such as the double-counting of carbon offsets and how the projects they support are tackling a broader range of environmental and social challenges being raised regularly.

Moreover, some other issues have been found related to carbon offsetting and carbon neutrality as for example:

- Non additionality of the carbon credits;
- Companies that claim to be carbon neutral just by buying carbon credits (100% offsetting);
- Companies that make small reductions and publicize them a lot;
- Companies that claim to be carbon neutral just on some scopes of their Carbon footprint (e.g. they claim to be carbon neutral scope 1–2 but the majority of their emissions are in scope 3).

4.1.1 Greenwashing and Companies' Sustainability Reporting

Another set of problems is related to how companies report and disclose carbon neutrality and offsetting information in their website, communication material or sustainability reports.

A **sustainability report** is a report published by a company or organization about the economic, environmental and social impacts caused by its everyday activities. A sustainability report also presents the organization's values and governance model and demonstrates the link between its strategy and its commitment to a sustainable global economy. Sustainability reporting can help organizations to measure, understand and communicate their economic, environmental, social and governance performance, and then set goals, and manage change more effectively. A sustainability report is the key platform for communicating sustainability performance and impacts – whether positive or negative(Global Reporting Initiative 2013).

GRI Standards: There is no universal standard for reporting CSR. The most commonly used reporting guidelines are produced by the Global Reporting Initiative (GRI). Global Reporting Initiative (GRI) Standards were developed to provide credibility and meaningful content for sustainability reports. The GRI was jointly founded by the Coalition for Environmentally Responsible Economies (CERES) and the United Nations Environment Program who are both still involved with the GRI (Global Reporting Initiative 2016).

The problem of greenwashing: When Corporate Social Responsibility (CSR) reports enhance positive social and environmental information which could contribute to misleading and biased reports the phenomenon greenwashing emerges (Mahoney et al. 2013). Greenwashing is defined as intentionally misleading or deceiving consumers with false claims about an entity´s environmental practice and impact (Nyilasy et al. 2014). Entities could use greenwashing to deceive their stakeholders and there is no formal protection against it, this due to the lack of regulation and punishment (Delmas and Burbano 2011).

The incentive behind greenwashing varies between different entities and situations, one reason could be when an entity has a damaged reputation and tries through the use of greenwashing to repair the public reputation and shape the perception of the entity. Greenwashing could also be done in different ways, in the CSR report, through marketing et cetera, and it is sometimes imbedded in rhetoric, which make it difficult to spot.

There are different drivers of greenwashing which could be divided into three categories; **external, organizational and individual**. The **external drivers** include the pressures from non-market actors (NGOs and regulators) and from market actors (consumers, investors and competitors). **Organizational drivers** include among others ethical perspective and entity incentive structure et cetera. **Individual drivers** include narrow decision framing, individual decision making and optimal bias (Delmas and Burbano 2011).

There are different types of greenwashing in sustainability reporting (Petersson and Dzafic 2016):

- **Reading Ease Manipulation** (making the accounting narratives more difficult to read, hide negative news);
- **Rhetorical Manipulation** (using persuasive language try to convince the reader of a better result);
- **Thematic Manipulation** (news content/tone) (overstate good news and understate bad news);
- **Visual and Structural Manipulation** (emphasis) (overstate the good news and position it first in the document. It also entails the choice of color, size, hiding bad news within the document and the use of repetition);
- **Performance Comparison** (choice of benchmark numbers and years to compare with the recent numbers);
- **Choice of Earnings Number** (selectivity) (selection of numbers that entities choose to include the best numbers in the narratives);
- **Attribution of Performance** (entity takes credit for the positive outcomes and blames external factors for negative outcomes).

4.1.2 *Sustainability Reporting: What Investors Are Asking*

A recent survey by McKinsey "More than values: The value-based sustainability reporting that investors want" shows that investors say they cannot readily use companies' sustainability disclosures to inform investment decisions and advice accurately (Bernow et al. 2019). Results of the McKinsey report shows that there is a need to conform companies' sustainability disclosures to shared standards in the way their financial disclosures must in order to perform comparisons among companies. Additionally, both executivesand investors strongly support to reduce the number of standards for sustainability reporting and investors also support legal mandates requiring companies to issue sustainability reports (Fig. 4.2).

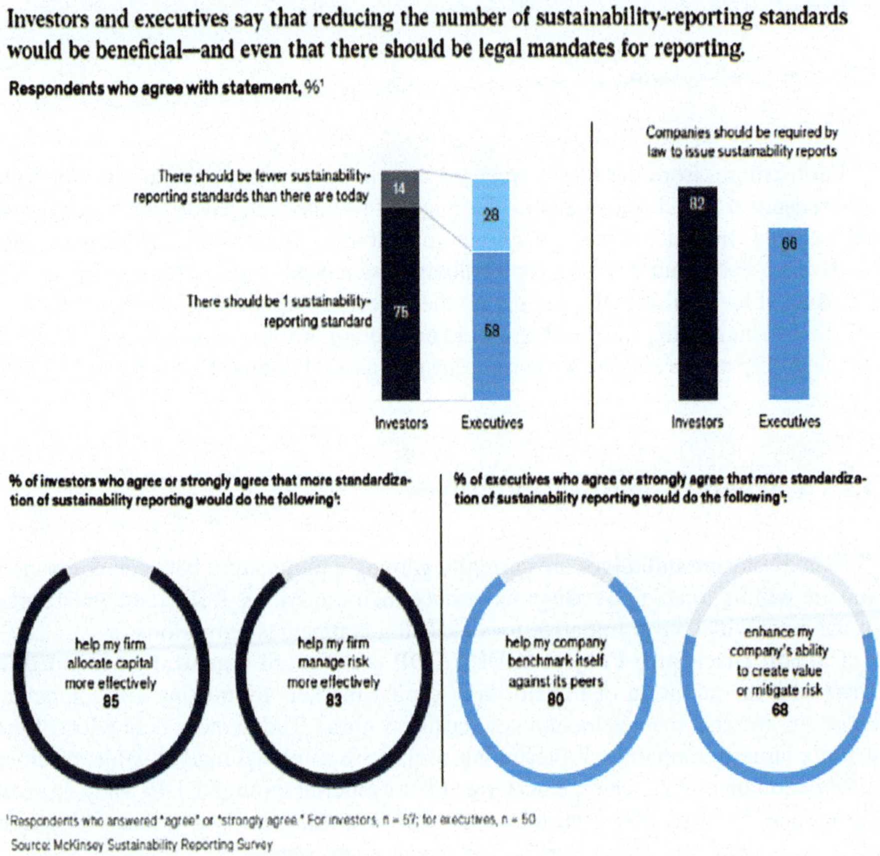

Source: (Bernow et al., 2019).

Fig. 4.2 McKinsey survey' results -Part 1
Source: Bernow et al. (2019)

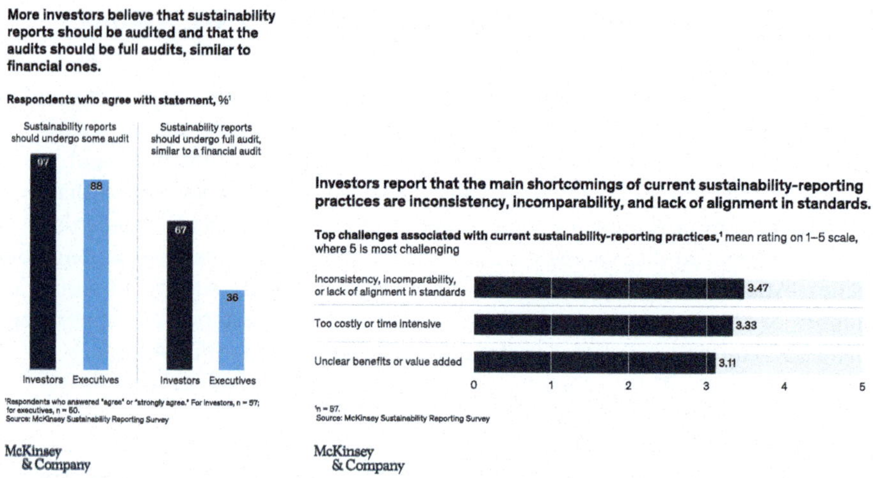

Fig. 4.3 McKinsey survey' results – Part 2
Source: Bernow et al. (2019)

Furthermore, from the report emerged that investors want companies to provide more sustainability disclosures that are material to financial performance, consistent and reliable. In fact, inconsistency, incomparability, or lack of alignment in standards have been found as the top challenge associated with current sustainability reporting (Fig. 4.3). Finally, almost all the investors McKinsey surveyed—97%—said that sustainability disclosures should be audited in some way, and 67% said that sustainability audits should be as rigorous as financial audits (Fig. 4.3).

4.2 Carbon Disclosure Initiatives

Carbon disclosure initiatives are currently gaining a momentum between companies that are willing to act to measure and reduce their emissions. Below are listed some of the most widespread initiative for disclosing companies' emissions.

Carbon Disclosure Project (CDP): CDP is nonprofit organization founded in 2000 with the ambition of transforming capital markets by making environmental reporting and risk management a new business norm. Today more than 6300 of the world's largest companies, representing some 55% of global market value, disclose information on climate change, water and deforestation through CDP. They do so at the request of over 650 investors with US$87 trillion in assets, and 115 major purchasers with over US$3.3 trillion in procurement spend.

Carbon Action Initiative Carbon Action is an investor-led initiative which shows how companies in investment portfolios are managing carbon emissions and energy efficiency. Over 300 investors with US$25 trillion in assets under management ask the world's highest emitting companies to take three specific actions in response to climate change: Make emissions reductions (year-on-year); Publicly disclose emission reduction targets; Make ROI-positive investments in projects. In 2016, the Carbon Action request went to over 1300 of the highest emitters. CDP launched a new research series at the beginning of 2015, taking a sector by sector approach.

Climate Disclosure Standards Board The Climate Disclosure Standards Board (CDSB) is a non-profit organization working to provide material information for investors and financial markets through the integration of climate change-related information into mainstream financial reporting. To this end, CDSB acts as a forum for collaboration on how existing standards and practices can be used to link financial and climate change-related information using its Framework for reporting environmental information, natural capital and associated business impacts. CDSB's research that was commissioned to prepare for the UK Government's Department of Environment and Rural Affairs (DEFRA). During the 2007 World Economic Forum in Davos, CDSB was formed to bridge the gap between the current uncoordinated approaches of reporting by having developed and promoting the Framework for reporting environmental information, natural capital and associated business impacts. CDP acts as a secretariat to CDSB.

Task Force on Climate-Related Financial Disclosure (TCFD) – G20 In April 2015, the G20 Finance Ministers and Central Bank Governors asked the **Financial Stability Board** to convene public- and private-sector participants and review how the financial sector can take account of climate-related issues. As part of its review, the **Financial Stability Board** identified the need for better information to support informed investment, lending, and insurance underwriting decisions and improve understanding and analysis of climate-related risks. To help identify the information needed by investors, lenders, and insurance underwriters to appropriately assess and price climate-related risks and opportunities, the Financial Stability Board established an industry-led task force: the **Task Force on Climate-related Financial Disclosures** (Task Force or TCFD). The Task Force was asked to develop voluntary, consistent climate-related financial disclosures that would be useful to investors, lenders, and insurance underwriters in understanding material risks. The 29-member Task Force is global; and its members were selected by the Financial Stability Board and come from various organizations, including large banks, insurance companies, asset managers, pension funds, large non-financial companies, accounting and consulting firms, and credit rating agencies.

Science Based Targets (SBTs) The Science Based Targets initiative mobilizes companies to set science-based targets and boost their competitive advantage in the transition to the low-carbon economy. It is a collaboration between CDP, the United Nations Global Compact, World Resources Institute (WRI) and the Worldwide Fund for Nature (WWF) and one of the We Mean Business Coalition commitments. 741 companies taking action to implement greenhouse gases targets in line with a 2 degrees world. The "Science Based Targets"initiative was created to help companies set quantified GHG emission reduction targets in line with the level of decarbonization required to keep the global temperature rise of Earth below 2 ° C by 2100. In order for target(s) to be recognized by the Science Based Targets initiative (SBTi) Companies should follow the GHG Protocol Corporate Standard, Scope 2 Guidance, and Corporate Value Chain (Scope 3) Accounting and Reporting Standard. The targets must cover company-wide scope 1 and scope 2 emissions, as defined by the GHG Protocol Corporate Standard. Companies may exclude up to 5% of scope 1 and scope 2 emissions combined in the boundary of the inventory and target. The targets must cover all relevant GHGs as required per the GHG Protocol Corporate Standard.

References

Bernow S, Godsall J, Klempner B, Merten C (2019) More than values: the value-based sustainability reporting that investors want. McKinsey Co.

Birkenberg A, Birner R (2018) The world's first carbon neutral coffee: lessons on certification and innovation from a pioneer case in Costa Rica. J Clean Prod 189:485–501. https://doi.org/10.1016/j.jclepro.2018.03.226

Bunning J, Beattie C, Rauland V, Newman P (2013) Low-carbon sustainable precincts: an Australian perspective. Sustain. https://doi.org/10.3390/su5062305

Delmas MA, Burbano VC (2011) The drivers of greenwashing. Calif Manag Rev. https://doi.org/10.1525/cmr.2011.54.1.64

Global Reporting Initiative (2013) Sustainability topics for sectors: what do stakeholders want to know? Res Dev Ser

Global Reporting Initiative (2016) Gri 101: foundation 2016 101. GRI Standards

Horiuchi R, Schuchard R, Shea L, Townsend S (2009) Understanding and preventing greenwash: a business guide. Futerra Sustain Commun:1–41

Hyams K, Fawcett T (2013) The ethics of carbon offsetting. Wiley Interdiscip Rev Clim Chang. https://doi.org/10.1002/wcc.207

Mahoney LS, Thorne L, Cecil L, LaGore W (2013) A research note on standalone corporate social responsibility reports: signaling or greenwashing? Crit Perspect Account. https://doi.org/10.1016/j.cpa.2012.09.008

Nerlich B, Koteyko N (2009) Compounds, creativity and complexity in climate change communication: The case of "carbon indulgences". Glob Environ Chang. https://doi.org/10.1016/j.gloenvcha.2009.03.001

Nyilasy G, Gangadharbatla H, Paladino A (2014) Perceived greenwashing: the interactive effects of green advertising and corporate environmental performance on consumer reactions. J Bus Ethics. https://doi.org/10.1007/s10551-013-1944-3

Petersson A, Dzafic J (2016) Greenwashing in CSR reports. Jönköping University

Wang HJ (2017) Determinants of consumers' purchase behaviour towards green brands. Serv Ind J 37:896–918. https://doi.org/10.1080/02642069.2017.1365140

Chapter 5
The Governance of the Carbon Neutrality System: Is Carbon Neutrality a Tool for the Low-Carbon Transition?

The chapter describes the carbon neutrality process and the different phases to achieve a carbon neutral status: measure, reduce, inset, offset, and validation and certification. Particularly, existing verification, certification, and labeling mechanisms for each of these phases have been presented. The chapter has been enriched with numerous company case studies to provide a practical perspective on the process.

5.1 Carbon Neutrality: A Nudging Strategy Towards the Decarbonization

The carbon neutrality path is a step-by-step process. Particularly, the main phases of the Carbon neutrality process are:

1. **Calculate the company' Carbon Footprint:** Carbon Footprint (CF) is a means for measuring, managing and communicating greenhouse gas (GHG) emissions related to organizations, goods and services. A carbon footprint is based on a life cycle assessment (LCA) but focuses on a single issue which is global warming.
2. **Plan the Emissions reduction and develop a carbon reduction plan:** Develop initiatives to reduce GHG emissions within the entity. This step starts with a commitment (target) implemented with an action plan.
3. **Offset emissions:** Offsetting projects are specific project or activity designed to achieve GHG emission reductions, storage of carbon, or enhancement of GHG removals from the atmosphere. A Carbon Credit is a unit generated by an offsetting project. When traded, is referred to as a carbon 'credit' - representing one metric tonne of CO2 equivalent - on either the voluntary or compliance carbon market.
4. **Insetting emissions:** entails reducing GHG emissions or sequestering carbon through an activity linked to the supply chain of a given actor or an activity in its

A. Acampora et al., *Carbon Neutrality in the Agri-food Sector*,
https://doi.org/10.1007/978-3-030-88048-4_5

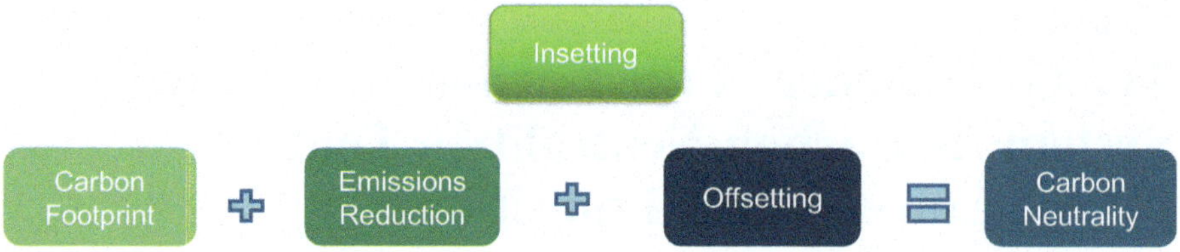

Fig. 5.1 Carbon Neutrality Guiding Scheme
Source: Authors' own elaboration

direct sphere of influence. These activities may or may not be monetized through the formal carbon markets.

5. **Achieve Carbon neutrality:** Carbon neutrality has been defined as a condition in which during a specified period there has been no net increase in the global emission of greenhouse gases to the atmosphere as a result of the greenhouse gas emissions associated with the subject during the same period.
6. **Carbon neutrality verification and certification:** Companies can choose to verify and certify all the carbon neutrality path or just some of the phases. Independent verification ensures transparency and rigor around carbon neutrality certifications and claims.
7. **Carbon neutrality labels:** Finally, companies can choose to have carbon neutral labels that are additional to certification. Companies can both developing their own label and using labels provided by different service providers (Fig. 5.1).

5.2 Different Approaches to Use the Carbon Footprint

Using different case studies in this Section we will explain what strategies companies can undertake starting from the Carbon Footprint calculation. In fact, a lot of confusion surrounded the different terms and strategies related to companies low-carbon transition.

In fact, companies nowadays companies have mainly followed the following strategies:

1. Carbon quantification/measurement
2. Carbon reduction
3. Carbon neutrality
4. Climate/Carbon positive

5.2.1 Carbon quantification /Measurement

In 2009 Tesco, Britain's largest retailer started a project to use carbon labels to more than 100 of its own-brand products, including pasta, milk, orange juice and toilet paper. In 2013 due to scarce level uptake by consumers the project was shut down (Fig. 5.2).

Casino launched in 2008 The Casino Carbon Index which is a measure of the greenhouse gases emitted during 5 key stages in the life cycle of each Casino brand product: production; manufacture; transport, from the field to the Casino warehouses; packaging, from raw material extraction to recycling; and distribution, from the Casino warehouses to the consumer's home. The carbon footprint is expressed in terms of grams of CO2 generated per 100 g of product. The Carbon Index is only valid in Metropolitan France.

5.2.2 Carbon Reduction

Project Gigaton is a Walmart initiative to avoid one billion metric tons (a gigaton) of greenhouse gases from the global value chain by 2030. Through Project Gigaton, suppliers can take their sustainability efforts to the next level through goal setting and get credit from Walmart for the progress you make. Since the program was introduced in 2017, hundreds of Walmart suppliers have gotten on board by committing to reduce emissions.

Heineken in 2018 announced its 'Drop the C' program for renewable energy. With 'Drop the C' the company aims to grow its share of renewable thermal energy and electricity in production from the current level of 14% to 70% by 2030 (Fig. 5.3).

Fig. 5.2 Tesco and Casino logos

Fig. 5.3 Walmart and Heineken logos

5.2.3 Carbon Neutrality

Since 2008, they have been compensating their full carbon footprint, which includes GHG Protocol Scopes 1, 2 and 3 from their complete product line. Every year, to offset their carbon emissions, they work with PUR Projet and our cacao producing partners at ACOPAGRO to reforest the San Martín region of Peru. This area suffered severe deforestation in the 1980s as it became a red zone for coca production. Alter Eco farmers, now cultivating high-quality cacao, strategically plant native and high-value trees within their cacao fields, to naturally sequester carbon and maintain the rainy microclimate so necessary for successful cacao cultivation (Fig. 5.4).

5.2.4 Climate/Carbon Positive

Unilever in 2015 set the goal of becoming carbon positive by 2030. Since 2015, they've been making progress in investing in clean technologies, increasing their energy efficiency and in switching to renewable energy sources. They extended the scope of our carbon positive ambition in 2017, from manufacturing to all operations, including distribution centers, research laboratories, marketing and sales offices and data centers.

Max Burgers launched the world's first "climate positive" menu in June 2018. Max Burgers ensures that each item on its menu is "climate positive", taking into account all emissions from the "farmers land to the guests' hand", while even including the customer's journey back and forth to the restaurant. This has required extensive efforts by the company to measure its entire footprint. Each item on Max Burger's menu includes a CO_2e label to empower customers to better understand the climate impact of food and guide them towards options with a lower carbon footprint. The initiative builds on the company's 2016 launch of plant based "green burgers", which have a fraction of the climate impact of beef burgers (Fig. 5.5).

Fig. 5.4 Alter Eco logo

Fig. 5.5 Max Burger and Unilever logos

Moreover, looking at different companies' pathways today we can find different carbon neutrality claims. There are companies that communicated their carbon neutrality commitment:

- **Carbon Neutral Commitment (by 20xx)** (Fig. 5.6)

There are companies that already carbon neutral and that issued their Carbon neutrality statement.

- **Carbon Neutral Statement (already done)** (Fig. 5.7)

Then, there are companies that once achieved the carbon neutral state have chosen to use a specific carbon neutral label.

- **Carbon Neutral Labels** (Fig. 5.8)

Fig. 5.6 Logos of companies that have declared a Carbon Neutral Commitment

Fig. 5.7 Logos of companies that have issued a Carbon Neutrality Statement

Fig. 5.8 Logos of companies that have adopted a Carbon Neutral Label.

5.3 Carbon Offsets

A carbon 'offset' is essentially a measure of GHG emissions reduction or carbon sequestered, relative to an initial baseline level. An offset represents the reduction, removal or avoidance of greenhouse gas emissions, measured in tons of CO2 equivalent (tCO2e), from a sector/region not subject to an emissions cap (International Emissions Trading Association 2019).

According to International Emissions Trading Association (2019) offsetting must demonstrate actual emission reductions compared to what would have otherwise happened, ensure emissions are not simply released at a later date, or are displaced elsewhere. Some of the consistent essential criteria used in existing greenhouse gas offset programs include:

- **Real:** offsets must represent real emission reductions that have already occurred (ie, the reduction is not projected to occur in the future)
- **Additional:** offsets must represent emission reductions that are in addition to what would have occurred otherwise
- **Permanent:** offsets must represent emission reductions that are non-reversible, or must typically be sequestered for X number of years in the case of carbon bio-sequestration projects
- **Verifiable:** sufficient data quantity and quality must be available to ensure emission reductions can be verified by an independent auditor against an established protocol or methodology
- **Quantifiable:** emission reductions must be reliably measured or estimated, and capable of being quantified
- **Enforceable:** offset ownership is undisputed and enforcement mechanisms exist to ensure that all programme rules are followed and the market's environmental integrity is maintained

When traded a Carbon offset, is referred to as a **carbon 'credit'** - representing one metric ton of CO2 equivalent - on either the voluntary or compliance carbon market. Carbon credits have been defined by the ICROA as a "unit of carbon dioxide equivalent which has been reduced, avoided or sequestered by a carbon reduction project and is a tradeable commodity" (ICROA, 2016).

- **Compliance markets** are created and regulated by mandatory regional, national, and international carbon reduction regimes, such as the Kyoto Protocol and the European Union's Emissions Trading Scheme.
- **Voluntary offset markets** function outside of the compliance markets and enable companies and individuals to purchase carbon offsets on a voluntary basis.

The voluntary offset market enabled private companies to buy carbon credits on a voluntary basis, most often as a tool for corporate social responsibility and to improve their public image.

These credits are bought and sold in the so-called "voluntary" market, which is not backed by any government standard or mandatory goals, but rather based on specific organizations certifying that emission reductions have environmental integrity (called "greenhouse gas (GHG) programs"). The entire market, therefore, rests on the relationship of trust between buyers and the GHG programs, and the claim that the credits sold on the market truly contribute to reducing emissions (Carbon Market Watch 2019).

5.3.1 Carbon Offset Standards

Launched in May 2006 by WWF-UK (a nonprofit foundation). It is a simplified version of the CDM Gold Standard, using the same basic methodologies. Only available for projects in developing countries. They are focused on renewable energy and energy efficient projects with strong sustainable development benefits. Eligible sectors are RE, EE, waste, land use and forests (afforestation, reforestation and agriculture) and water (supply, purification and conservation) (Gold Standard, 2018f). Under this standard, more than 550 registered projects have achieved emission reductions of about 78 million tCO2eq in the period 2008–2017 (Gold Standard 2018a). The foundation has launched a new generation of standards with the dual objective of pursuing climate action and the fulfilment of SDGs, Gold Standard for the Global Goals (GS4GG) in 2017 (Gold Standard, 2018g) (Fig. 5.9).

Verra (formerly known as the VCS) is a not-for profit organization founded in 2005, serving as a secretariat to various standards. Verra's flagship is the Verified Carbon Standard (VCS) with CDM-like MRV requirements. It issues Verified Carbon Units (VCUs). In terms of volume, it is the largest voluntary standard in the world, having certified reductions of more than 200 million tCO2eq from more than 1300 projects since 2006 in a wide range of sectors. So far, more than 100 projects covering more than ten million hectares of land have been validated, with 40 having achieved full verification (Verra 2019). This standard was developed by the Climate Group and International Emissions Trading Association (IETA). It provides real, quantifiable, additional and permanent projects- based emission reductions. Credits are managed through registries to register, transfer and retire Voluntary Carbon Units (VCUs) (Fig. 5.10).

In 2015, the UNFCCC created a dedicated website where organizations, companies, but also private persons are able to offset their footprint with the aim of

Fig. 5.9 The Gold Standard logo

Fig. 5.10 The Verra logo

Fig. 5.11 The UNFCCC logo

facilitating everyone's participation in the process of promoting sustainability on a voluntary basis. This platform features UNFCCC certified projects that reduce, avoid or remove greenhouse gas emissions from the atmosphere. The projects are implemented in developing countries and are rewarded with Certified Emission Reductions (CERs), a type of carbon offset measured in tons of CO2 equivalent. CERs that are issued from trustworthy climate-friendly projects called Clean Development Mechanism (CDM) projects. CDM projects take place in developing countries and contribute to their sustainable development. Each project goes through a strict and thorough vetting process. The CDM process involves variety of stakeholders such as the project participants who own the projects, host-country national authorities who oversee national implementation, independent auditors know as the Designated Operational Entities, the UNFCCC CDM Executive Board and its secretariat. At a higher level, all CDM work is coordinated and directed by the UNFCCC Conference of the Parties to the Kyoto Protocol (CMP), the ultimate body responsible for the implementation of the Kyoto Protocol where all member states take collective decisions (Fig. 5.11).

Plan Vivo is a program designed for community- based forest management and agroforestry payments for ecosystem services projects. The system was created over a decade ago by the Edinburgh Center for Carbon Management and is now developed and overseen by a Scottish charity, the Plan Vivo Foundation. There are currently five fully operational Plan Vivo projects in Mexico, Uganda, Mozambique, Tanzania and Nicaragua and several upcoming projects in developing countries including Malawi, Cameroon, Ethiopia, and Nepal. Plan Vivo maintains a listing of projects on its website and lists credits (Plan Vivo Certificates) on the Markit Environmental Registry. Currently the activities that are eligible to generate Plan Vivo Certificates are afforestation and reforestation, agroforestry, avoided deforestation, forest conservation and restoration. Plan Vivo projects have engaged tens of thousands of smallholder farmers to plant multipurpose trees on their own land, and an even greater number of rural community participants involved in restoration and protection activities as well as livelihood development initiatives. These deal with land-use change and forestry predominantly, channeling some $17.5 million USD to projects targeting rural communities in developing countries. As of March 2017, these projects have collectively established, and are helping conserve more than 138,000 hectares of forest (Plan Vivo 2019) (Fig. 5.12).

SOCIALCARBON is a Standard developed by the Ecologica Institute that certifies carbon reduction projects for their contributions to sustainable development. Six aspects of project sustainability are individually measured using the SOCIALCARBON hexagon: carbon and biodiversity as well as social, financial, human and natural components. As an additional standard for co-benefits,

Fig. 5.12 The Plan Vivo logo

Fig. 5.13 The Social Carbon logo

Fig. 5.14 The CCBS logo

SOCIALCARBON can be implemented alongside any other carbon accounting standard (e.g. VCS, CDM, CAR or others) and may be adapted to suit different types of projects, including hydropower plants, landfills, fuel switching, forestry and others. The theoretical framework of SOCIALCARBON is based on the Sustainable Livelihood Approach (SLA), a worldwide methodology used in planning new development activities and assessing the contribution that existing activities have made to sustaining livelihoods (Fig. 5.13).

Verra furthermore manages the Climate, Community and Biodiversity Standard (CCBS), certifying land-based climate change projects that pursue multiple benefits: improve livelihoods, create employment, protect traditional cultures and endangered species, help secure tenure to lands and resources, increase the resiliency of ecosystems and help to combat climate change. CCBS certification can be applied also to VCS projects. This standard has been developed by the Climate Community and Biodiversity Alliance. It is for land-based projects that can simultaneously deliver compelling climate biodiversity and community benefits. It uses methodologies of the intergovernmental panel on climate change good practice guidance (IPCC GPG) but can also use approved CDM methodologies for calculating carbon reductions/savings (Fig. 5.14).

Finally, we elaborated Table 5.1 to compare the main characteristic of the most diffused carbon offset standard.

Table 5.1 Comparisons of the main Carbon offset standards

Owner	Name	Logo	Year of creation	Objectives	What is certified?
Gold Standard	Gold Standard (GS)	Gold Standard Climate Security & Sustainable Development	2006	Certify carbon-offset projects	Voluntary Emission Reduction (VER's)
Verra	Verified Carbon Standard (VCS)	Verified Carbon Standard	2006	Certify carbon-offset projects	Verified Carbon Units (VCU's)
Plan VIVO	Plan Vivo Standard	Plan Vivo Improving livelihoods, restoring ecosystems	2007	Certify carbon-offset projects with focus on co-benefits	Plan Vivo Certificates (PVCs)
UN	UNFCCC Clean Development Mechanism (CDM)		2008	Certify carbon-offset projects	Certified Emission Reduction (CER'S)
Ecologica Institute	SocialCarbon	SOCIALCARBON®	2003	Certify social, environmental, economic benefits of carbon-offset projects	Social quality for Voluntary Emission Reduction (VER's)
CCBA (VCS assumed management of the CCB Program in November of 2014)	Climate, Community & Biodiversity (CCB) Standard	CCBA	2005	Certify social, environmental, economic benefits of carbon-offset projects	Biodiversity and Community quality for Verified Carbon Units (VCU's)

Source: Authors' own elaboration

5.3.2 *Voluntary Offset Market Trends*

According to the report of Forest Trends' Ecosystem Marketplace (2019) "Financing Emissions Reductions for the Future: State of the Voluntary Carbon Markets 2019" airlines, oil companies, and individuals are using voluntary carbon markets to achieve net reductions in greenhouse gas emissions at levels not seen in 7 years.

The document reported that the volume of transactions in voluntary carbon markets hit a 7-year high in 2018, identifying transactions for **98.4 million metric tons of carbon dioxide (MtCO2e),** with a market value **of $295.7 million**. This represents a 52.6% increase in volume and a 48.5% increase in value over 2016.

Furthermore, the Forest Trends' Ecosystem Marketplace (2019) develop a historical analysis on the voluntary offset market development. Table 5.2 presents the results of this analysis. We can see as the average price per metric ton is constantly decreasing by years and from 2014 is around 3$. Moreover, after being decreasing from 2007–2008, also in conjunction with global economic crisis, both market value and volume from 2018 started growing again (Table 5.2).

Moreover, has highlighted by Forest Trends' Ecosystem Marketplace (2019) last year growth was led by a growing awareness of "**nature-based solutions**" for climate resilience, which drove a **264% increase in volume of offsets generated through Forestry and Land Use activities** and made REDD+ the most popular offset type for the first time since 2015. This surge in volume continued to accelerate in 2019, according to market experts (Fig. 5.15).

Additionally, for what concern the scope of this report is interesting to note that Forest Trends' Ecosystem Marketplace (2019) in the last years (between 2017 and 2018) modified their categorization of project types and categories (Figs. 5.16 and 5.17) adding Agricultural project category and particularly the following project types:

Table 5.2 Voluntary offset market trends

Year	Volume ($MtCO_2e$)	Market Value ($M)	Average Price ($)
2006	31,6	111,3	4,1
2007	69,8	359	6,1
2008	134,5	790,2	7,34
2009	107,2	484,5	6,46
2010	131,4	444,1	6
2011	100	602,3	6,2
2012	102,8	529,8	5,87
2013	68	338,5	4,93
2014	76,8	298,5	3,8
2015	84,1	278,2	3,26
2016	64,5	199,1	3,1
2017	46,2	145,8	3,16
2018	98,4	295,7	3,01

Source: Authors' own elaboration on Forest Trends' Ecosystem Marketplace (2019) data

	2017			2018		
	VOLUME $MtCO_2e$	AVERAGE PRICE	VALUE	VOLUME $MtCO_2e$	AVERAGE PRICE	VALUE
FORESTRY AND LAND USE	16.6	\$3.4	\$63.4M	50.7	\$3.2	\$171.9M
RENEWABLE ENERGY	16.8	\$1.9	\$31.5M	23.8	\$1.7	\$40.9M
WASTE DISPOSAL	3.7	\$2.0	\$7.4M	4.5	\$2.2	\$10.0M
HOUSEHOLD DEVICES	2.3	\$5.0	\$11.8M	6.1	\$4.8	\$29.5M
CHEMICAL PROCESSES/ INDUSTRIAL MANUFACTURING	2.6	\$1.9	\$4.9M	2.5	\$3.1	\$7.9M
ENERGY EFFICIENCY/ FUEL SWITCHING	1.1	\$2.1	\$3.3M	2.8	\$2.8	\$7.8M
TRANSPORTATION	0.1	\$2.9	\$0.2M	0.3	\$1.7	\$0.5M

Notes: 2017 figures are based on 1,041 transactions for a total volume of 43.2 $MtCO_2e$. 2018 figures are based on 1,568 transactions for a total of 90.7 $MtCO_2e$. These figures do not include responses that didn't provide price data.

Fig. 5.15 Transacted Voluntary Carbon offset volume, value and weighted average price by project category, 2017 and 2018
Source: Forest Trends' Ecosystem Marketplace (2019)

- Fertilizer - N20
- Grassland/rangeland management
- Livestock methane
- No-till/low-till agriculture
- Rice cultivation/management
- Sustainable agricultural land management
- Other – Agriculture

5.4 Carbon Insetting

Carbon insetting entails reducing GHG emissions or sequestering carbon through an activity linked to the supply chain of a given actor or an activity in its direct sphere of influence (Phelan 2015). By investing in their own value chain, and enhancing the resilience of the chain as a whole to climate change - including the most vulnerable link in the chain, namely, smallholder farmers responsible for primary production - agri-food industry actors can create shared value and generate mutual benefits, while reducing their carbon footprint and the GHG emissions associated with the goods produced (Phelan 2015).

Figures 5.18 illustrate the differences between carbon offsetting and insetting. In the case of offsetting the emissions and reductions are discrete activities and there is no interaction between the parties except a financial transaction. In the case of

Carbon Offset Project Types and Categories (pre-2018)

Project Category	Project Type
Energy Efficiency/Fuel Switching	Energy efficiency - community-focused (targeting individuals/communities/housing/campuses)
	Energy efficiency - industrial-focused (targeting corporations/industrial processes)
	Fuel switching
	Waste heat recovery
Forestry and Land Use	Afforestation/reforestation
	Agro-forestry
	Grassland/rangeland management
	Improved forest management
	No-till/low-till agriculture
	REDD - Avoided planned deforestation
	REDD - Avoided unplanned deforestation
	Rice cultivation/management
	Soil carbon
	Sustainable agricultural land management
	Urban forestry
	Wetland restoration/management
Gases	N20
	Ozone-depleting substances (Article 5)
	Ozone-depleting substances (US based)
Household Devices	Clean cookstove distribution
	Water purification device distribution
Methane	Coal mine methane
	Landfill methane
	Livestock methane
	Waste water methane
Renewable Energy	Biogas
	Biomass/biochar
	Geothermal
	Large hydro
	Run-of-river hydro
	Solar
	Wind
Other	Not specified
	Other

Fig. 5.16 Forest Trends' Ecosystem Marketplace Carbon offset project types and categories, pre-2018
Source: Forest Trends' Ecosystem Marketplace (2019)

insetting there is an exploration and partnership with various stakeholders to identify emission reduction opportunities (Tipper et al. 2009).

The term insetting was initially presented in 2009 to refer to the direct investment of a company within its own value chain (up- and down-stream) in order to reduce its carbon footprint (Davies 2016). According to Davies (2016) insetting can configurate in two ways:

- **Upstream carbon insetting** - where organizations seek to reduce emissions or sequester carbon using upstream suppliers;

Carbon Offset Project Types and Categories (2018)*

Project Category	Project Type
Agriculture	Fertilizer - N20
	Grassland/rangeland management
	Livestock methane
	No-till/low-till agriculture
	Rice cultivation/management
	Sustainable agricultural land management
	Other - Agriculture
Chemical Processes/Industrial Manufacturing	Nitric Acid
	Ozone-depleting substances (Article 5)
	Ozone-depleting substances (US based)
	Carbon capture and storage
	Coal mine methane
	Other - Chemical Processes/Industrial Manufacturing
Energy Efficiency/Fuel Switching	Energy efficiency - community-focused (targeting individuals, communities, etc.)
	Energy efficiency - industrial-focused (targeting corporations)
	Fuel switching
	Waste heat recovery
	Other - Energy Efficiency/Fuel Switching
Forestry and Land Use	Afforestation/reforestation
	Agro-forestry
	Avoided conversion
	Improved forest management
	REDD - Avoided planned deforestation
	REDD - Avoided unplanned deforestation
	Soil carbon
	Urban forestry
	Wetland restoration/management
	Other - Forestry and land use
Household Devices	Clean cookstove distribution
	Water purification device distribution
	Other - Household Devices
Renewable Energy	Biogas
	Biomass/biochar
	Geothermal
	Large hydro
	Run-of-river hydro
	Solar
	Wind
	Other - Renewable Energy
Transportation	Transportation - private (cars/trucks)
	Transportation - public (bikes/public transit)
	Other - Transportation
Waste Disposal	Landfill methane
	Waste water methane
	Other - Waste Disposal

***As defined by Ecosystem Marketplace. Please note that we modified our categorization of project types and categories between 2017 and 2018.**

Fig. 5.17 Forest Trends' Ecosystem Marketplace Carbon offset project types and categories, in 2018
Source: Forest Trends' Ecosystem Marketplace (2019)

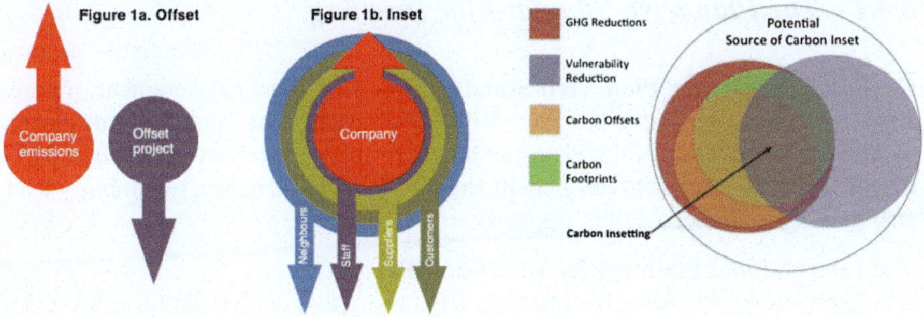

Fig. 5.18 Offsetting and Insetting
Source: Tipper et al. (2009)

- **Broader scope insetting** - including upstream and downstream as well as other materials besides carbon (reduced water or electricity usage, community development projects, reduced packaging material, etc.)

In the Davies (2016) research commissioned by ICROA two main definitions of insetting has been provided:

- **1° working definition:** The company invests in the development of a carbon offset project within its own supply chain and purchases all generated carbon credits to offset its operational emissions.
- **2° working definition:** The company invests in any activity within its supply chain that generates environmental, social and/or economic value for the supplier and company.

Several working groups such as the International Carbon Reduction and Offset Alliance (ICROA) are seeking to come to consensus in partitioning the idea of insetting into these two definitions, identify best practices and ultimately standardized methods for certifying insets (Climate Smart Group 2017).

Moreover the main Carbon Insetting definitions are listed below:

- **The Plan Vivo definition:** «Carbon insetting entails **reducing GHG emissions** or sequestering carbon through an activity **linked to the supply chain** of a given actor or an activity in its direct sphere of influence. These activities **may or may not be monetized through the formal carbon markets**» (Plan Vivo 2019)
- **The ICROA definition:** «The development of a **carbon offset** project within a **company's own supply chain** and supply chain communities»(Davies 2016)
- **The International Insetting platform (IPI) definition**: «Insetting represents the **actions** taken by an organization to **fight climate change** within its **own value chain** in a manner which generates **multiple positive sustainable impacts** » (International Insetting Platform 2017)

5.4.1 The Plan Vivo Standard for Insetting

Certification under the **Plan Vivo Standard** can enable the measurement of emissions reductions and generate Plan Vivo Certificates. The focus on sustainable livelihoods at the local level will ensure higher resilience to adverse climate impacts. Certification also demonstrates benefit sharing, good governance, community leadership, and positive ecosystem impacts.

The International Platform for Insetting (IPI)

The International Platform for Insetting (IPI) is a collaborative platform involving companies, academics, project developers, project operators, certification bodies, etc. aiming at giving visibility and promoting Insetting practices worldwide. It was founded as a non-profit organization in December 2013 in Paris by PUR Projet and Adaptogether. AccorHotels, Chanel and Nespresso France joined as corporate Founding members. The objective of the platform is to facilitate the integration of Insetting practices in the business world. IPI serves as a multi-stakeholder platform, working to better define Insetting, facilitate collaboration on common projects and to promote best practices. IPI further adds value for businesses and their Insetting Program and Projects by allowing them to be certified under the Insetting Standard by third parties and registered in a blockchain registry. The International Insetting Platform developed the **Insetting Program Standard (IPS)** that applies to organizations engaged in an Insetting Program consisting of at least one, or several, Insetting Projects. The aim of the Standard is to give guidelines, credibility and assurance on the Insetting strategies and Programs enforced by IPI Members. Broader than that, its ambitions are to set quality criteria for Insetting globally and help all players engaged in Insetting Programs progress towards excellence (Fig. 5.19).

However, according to the authors these definitions are generic and doesn't take into account the carbon neutral process of the companies. In this sense we develop two different conceptualization of insetting that are applicable if companies are developing a carbon neutral strategy according to the specifications of the PAS 2060:

Fig. 5.19 International Insetting Platform Logo and members
Source: International Insetting Platform (2017)

- If the company is developing the Carbon Neutrality in **scope 1 and 2** then the supply-chain is **OUTSIDE,** the scope of the Carbon Neutrality and insetting has the carbon credits valence; they can buy carbon credits on their supply-chain;
- If the company is developing Carbon Neutrality in **scope 1, 2, 3** then the supply-chain is **INSIDE** the scope of the Carbon Neutrality and insetting has reduction valence; they can develop emissions reduction projects in their supply-chain and have them accounted as reduction.

5.4.2 Insetting Case Studies

Nespresso is using Insetting to increase coffee yields, reduce input costs and diversify revenues for coffee farmers in Colombia and Guatemala. In partnership with Pur Projet, agroforestry systems are being implemented including shade trees, boundary planting, woodlots and integrating with animal pasture. The objective is to regenerate coffee ecosystems and help farmers to adapt to climate change, by making their farms more resilient. All the projects are verified against the IALL standard (Insetting via Agroforestry at Landscape Level) developed by the IPI (International Insetting Platform), while the program is being certified against the IPS (Insetting Program Standard); then all the program aspects, from its inputs and commitments to the verified outcomes, are registered in the IPI blockchain registry (Fig. 5.20).

This food manufacturer is using Insetting to secure its supply chains, take steps to mitigate 65,000 tons of CO_2 equivalent and demonstrate its environmental and social commitments in a positive way. In partnership with Pur Projet and local communities, Ben & Jerry's is participating in the preservation of a forest of more than 300,000 hectares in the Andean Amazon of Peru, which delivers key environmental services to the organic and fairtrade farmers of cocoa and coconut (Fig. 5.21).

Chocolats Halba is climate-neutral since 2011. The carbon footprint of Chocolats Halba was for the first time the subject of a joint analysis with the non-profit myclimate foundation in 2008. According to the analysis, Chocolats Halba emitted

Fig. 5.20 Nespresso logo

Fig. 5.21 Ben & Jerry's logo

3387 tons of CO_2 in 2008. Reduction: Mainly on account of switching from conventional electricity to hydroelectric power, Chocolats Halba was able to cut the company's CO_2 emissions by 27% in 2010. Compensation: To achieve climate neutrality, the company has been offsetting all remaining corporate emissions directly in its own value chain since 2011. Chocolats Halba is working with its partner cooperatives on reforestation projects in which smallhold farmers plant hardwood trees in addition to producing Halba cocoa. The reforestation project in Peru has been certified for emissions trading. In 2015, the project in Honduras will be certified for emissions trading in accordance with the "Gold Standard"– an accreditation for offset projects that also promote sustainable development in the countries and support local communities. Chocolats Halba has engaged in carbon insetting through its cocoa supply chain. By analyzing, reducing and compensating GHG emissions and 'offsetting all remaining corporate emissions in its own value chain', the Swiss premium manufacturer of chocolate specialties for retail and industry has succeeded in achieving its carbon reduction goals and in doing so, also realized its goal of climate-neutral chocolate production (Fig. 5.22).

Finally, in Fig. 5.23 the main Insetting benefits have been summarized.

5.5 Targets and Scopes of the Companies' Carbon Neutrality Path

5.5.1 Targets of the Carbon Neutrality

Companies can follow different paths for carbon neutrality. Firstly, they can choose what are the target of their carbon neutrality strategy. Mainly, companies have chosen to dedicate their efforts to reach Carbon neutrality for:

- Brand;
- Products;
- Organization;
- Services
- Events.

To provide some meaningful examples we elaborated a case study for each of these targets.

Fig. 5.22 Chocolats Halba logo

Fig. 5.23 Insetting benefits
Source: https://www.purprojet.com/presentation-what-is-insetting/

- **Brand**

Scope of CN	Wasa Brand; Scope 1-2-3
Carbon Footprint Measurement	GHG protocol; ISO 14064; Third-party verification
CN standard	DNV – GL, third-party certification for CN PAS 2060:2014
Reduction strategy	**Reduction**: Energy Saving Programs, Green Logistic Projects, and purchasing 100% Renewable Origin (GO-Guarantee of Origin) electricity for manufacturing plants
Offset strategy and certification	**Offsetting:** Peruvian Madre de Dios REDD+ project, labelled by VCS and the CCBA, as an emblematic project; Indian solar project – multisite- also labelled by VCS.
Label	Co2 Compensated Private label

- **Products**

Scope of CN	Baby formula products (since 2013); Milk cereals and baby porridges (since 2014); Got milk (since 2016) Scope 1-2-3
Carbon Footprint Measurement	GHG protocol, measurement undertaken by a third-party measuring emissions from field to shelf
CN standard	n.a.
Reduction strategy	**Reduction:** Reducing material used in packaging; Increase use of recycled material; Reducing energy use, using renewable energy; Installing solar panels on logistics building; Optimizing transport through IT software
Offset strategy and certification	**Offsetting:** Offsetting through composting projects;
Label	Co2 Neutral Private label

- **Organizations**

Scope of CN	Scope 1-2
Carbon Footprint Measurement	Unclear; 2nd party verified by Carbonfootprint ltd according to ISO 14064 Part 3 (2006)
CN standard	Carbon Footprint standard; Carbon Neutral Organisation →Private standard; 2nd party verification
Reduction strategy	**Reduction:** they have committed to reduce their absolute Scope 1 and 2 greenhouse gas emissions 100% by 2030 from a 2017 base-year. Reduce Scope 3 greenhouse gas emissions from crop to cup to 50% per million units of products by 2030 from a 2017 base year (**Science Based Target validated**)
Offset strategy and certification	**Offsetting:** offset their total operational emissions, plus a few additional measures including our staff commuting. Their offset in 2018 was 2,294 tCO2e. Their VCS-certified offsets help to invest in forest conservation in the Amazon rainforest, one of the world's 'biodiversity hotspots'. Closer to home, they also plant trees in the South West of England– they'll be planting them in 2019.
Label	Carbonfootprint ltd
Additional information	
Additional information	DATA SNAPSHOT - CARBON FOOTPRINT (see table below)

DATA SNAPSHOT - CARBON FOOTPRINT

PUKKA'S OPERATIONAL CARBON FOOTPRINT (tCO2e)	FY16	FY17	FY 2018	% Change
**Scope 1: natural gas and refrigerant loss	22.95	21.28	8.99	-58%
**Scope 2: purchased electricity *market based	17.20	36.55	7.52	-79%
Scope 3:Crop to Cup (calculated every two years)	–	15,184	–	
			FY18	
Operational energy use (kWh)			367,340	
Operational water use (m3)			2,018	

- **Services**

Scope of CN	La Poste Group Scope 1-2-3
Carbon Footprint Measurement	Unclear
CN standard	N.A.
Reduction strategy	**Reduction:** reduce the carbon footprint of transport, buildings and digital services: Larger fleet of electric vehicles, 100% electric supply, Energy Management System, First Green Bond issue, 2025 carbon trajectory approved by Science Based Targets
Offset strategy and certification	**Offsetting:** Offsetting by buying credits certified by VCS and Gold Standard in third world countries and working with the French national forest ownership centre (CNPF) invested in innovative carbon sequestration projects in France. By taking part in several pilot forest management project.
Label	Co2 Neutral Private label

- **Events**

Scope of CN	Expo event; Direct and indirect emissions
Carbon Footprint Measurement	ISO 14064 For CO2 Inventory
CN standard	N.A.
Reduction strategy	**Reduction:** the construction of permanent buildings and improve performances in terms of energy efficiency, and the installation of LED lighting; the creation of a Digital Smart City; the purchase of green electricity; the implementation of a programme for goods reuse at the end of the Event and for maximising the recycling of building materials and non-reusable goods.
Offset strategy and certification	**Offsetting:** The purchase of carbon credits was carried out in 2015, for a total of 279,000 credits (offsetting an equivalent number of tCO2 eq) generated by 11 projects certified by recognized standards (CDM, Gold Standard, VCS, Social Carbon, ICROA and CCB) and chosen according to a reliable carbon offsetting strategy, coherent with the sustainability policy and the themes of the Exposition of Milan
Label	No label

5.5.2 Scope of the Carbon Neutrality

Then, companies have to choose the scope of their carbon neutrality strategy. As we already explained in Sect. 3.3.2, organizational scopes have been defined by the GHG protocol as Scope 1 (Direct emissions that result from activities within your organization's control), Scope 2 (Indirect emissions from any electricity, heat or stream you purchase and use) and Scope 3 (other indirect emissions from sources outside your direct control).

Figure 5.24 provides some examples of companies that are carbon neutral or that have committed to became carbon neutral according to the scope they choose to implement their strategy.

Fig. 5.24 Carbon neutral commitment of companies and brands
Source: Authors' own elaboration

Below, we provide some examples of different companies' strategies according to scope they choose and to the level of implementation (already carbon neutral or in the way to be carbon neutral).

- **Fully carbon neutral - Scope 1 + 2 + 3**

Company	Dilmah
Scope of CN	Scope 1+2+3
Year project started	2013
Carbon Footprint Measurement	GHG protocol
CN standard	n.a.
Reduction strategy	**Reduction:** Energy efficiency in plant operations; Energy efficiency in transportation; Substitution of fossil energy with renewable energy; Recycling of waste (Total reduction 14%)
Offset strategy and certification	**Offsetting:** offsetting UN approved Cleaner Development Mechanisms (CDM).
Label	Private Label

- **In the way to be carbon neutral - Scope 1 + 2 + 3**

Company	Sky
Scope of CN	Since 2006 Carbon neutral Scope 1+2- by 2030 net zero carbon by 2030 across their value chain (Scope 1 +2+3)
Year project started	2006
Carbon Footprint Measurement	Greenhouse Gas Protocol - A Corporate Accounting and Reporting Standard -> (Scope 1+2+3)
CN standard	n.a

(continued)

Reduction strategy	**Reduction target:** The Science Based Targets Initiative is checking their net zero targets (2030); **Reduction strategy**: Since 2006 ->reduce emissions from transport and waste disposal and improve energy efficiency. By 2030 -> scope 1 and 2 - halve carbon intensity; commitment to use 100% renewable energy and invest in on-site renewable; transition 5,000 vehicles to zero emissions; sustainable product design and deliver; increase energy efficiency measures across new and existing buildings scope 3 - halve the CF from the use of products in customer homes by 2030; Work with their 11,000 suppliers to influence and partnerships to advocate for change through CDP Supply Chain; collaboration with manufacturing suppliers of our Sky Products such as supporting energy efficiency audits and enabling investment in on site renewables.
Offset strategy and certification	Since 2006-> natural carbon sinks such as mangroves restoration, and reforestation, but also, included renewable energy projects - carbon credits that have been verified to the Voluntary Carbon Standard. By 2030 -> committing that 100% 'carbon neutral' offsets - projects related to natural carbon sinks such as forests, mangroves and seagrass
Label	no

- **Carbon neutral - Scope 1 and 2**

Scope of CN	MARKS & SPENCER Scope 1+2
Year project started	2012
Carbon Footprint Measurement	GHG Corporate Reporting and Accounting Standard
CN standard	PAS 2060
Reduction strategy	**Reduction:** In June 2017 M&S published a Science Based Target initiative approved target of reduce absolute scope 1 and 2 greenhouse gas emissions 80% by 2030 below 2007 levels and has a longer-term vision to achieve 90% absolute emissions reductions by 2035, below 2007 levels. Marks & Spencer also commits to reduce scope 3emissions by 13.3 MtCO2e between 2017 and 2030
Offset strategy and certification	**Offsetting:** Offsetting with credits certified by VCS and Gold standard

(continued)

Label	Carbon Trust Label

- **In the way to be carbon neutral - Scope 1 and 2**

Scope of CN	Scope 1 +2 Breweries by 2030 Carlsberg
Year project started	2015
Carbon Footprint Measurement	Unclear is working with Carbon Trust
CN standard	n.a.
Reduction strategy	**Reduction:** Since 2015, they've reduced their relative carbon emissions by 20%, with 46% renewable electricity in 2018; they are working with SBT to set targets for 2030; by 2022 they will have ZERO carbon emissions at our breweries; 30% reduction in beer-in-hand carbon footprint; 50% reduction in carbon emissions at our breweries; 100% electricity from renewable sources at our breweries; ZERO coal at our breweries; 15% reduction in beer-in-hand carbon footprint; 100% low climate impact cooling
Offset strategy and certification	n.a.
Label	Carbon trust

5.6 Voluntary Standards and Carbon Neutrality

According to the International Organization of Standardization (ISO), standards can be defined as "*documented agreements containing technical specifications or other precise criteria to be used consistently as rules, guidelines or definitions, to ensure that materials, products, processes and services are fit for their purpose*" (International Organization for Standardization/ International Electrotechnical Commission (ISO/IEC) 1996, 2001).

According to the OCED report written by Rousset et al. (2015) standards can be:

- Voluntary: without any legal obligation.
- Mandatory: required or commanded by authority; obligatory, compulsory.

Voluntary standards have been defined as: «*a document approved by a recognized organization or entity, that provides, for common and repeated use, rules, guidelines or characteristics for products or related processes and production methods, with which compliance is* ***not mandatory*** *under international trade*

Enforcement	Ownership		
	Public		Private
	Governmental	Inter-governmental	
Voluntary (enforcement by certifications bodies)	Governmental standards established by a legislature and/or public regulator, including governments, supra-national and local governments	Inter-governmental standards established by international organisations as the result of an agreement between participating governments	Private standards established by standard-setting bodies and other non-governmental organisations, industry bodies, coalitions or individual firms
Mandatory (non-compliance is illegal; enforcement by criminal and administrative courts)	Technical regulations adopted by a legislature and/or public regulator, including governments, supra-national and local governments	N/A[1]	N/A[1]

1. When governments incorporate the technical requirements established by a standardising body or an international organisation as binding rules through their national legislation process, the documents acquire a legal position of domestic regulation.

Fig. 5.25 Comparison of standards and technical regulations
Source: Rousset et al. (2015)

rules. It may also include or deal exclusively with terminology, symbols, packaging, marking or labelling requirements as they apply to a product, process or production method." (International Trade Center 2019).

The OCED reported the following classification for standards (Fig. 5.25) (Rousset et al. 2015).

However, in literature there are both terminological and conceptual confusion with regard to the differences between private, public, voluntary, mandatory, and regulatory standard. Neither there are common and shared definitions and classification of the different types of standards.

For these reasons, we will use the classification provide by the International Centre for Trade and Sustainable Development (ICTSD) (2017). The ICTSD developed a differentiation between International Standards, Regulations and Private standards (Fig. 5.26):

- **International Standards:** Set by private–public partnerships, including CSOs, academia, regulators, and industry representatives. Compliance with relevant requirements is in principle voluntary, but can become mandatory, e.g. when standards are incorporated in regulations.
- **Regulations:** Set, adopted, and applied by the government directly or indirectly. May be based on standards, particularly "international", but also "private" standards. Compliance with relevant requirements is mandatory.
- **Private standards:** Set and operated by private companies, CSOs, or joint initiatives thereof. Compliance with requirements is voluntary, but it can become de facto mandatory when the standard setter, e.g. a large distributor, has a particularly dominant position in a given product or geographic market.

For what concern all the different elements that compose the carbon neutrality implementation process described in the guiding scheme (Fig. 5.27) there are available only voluntary standards. Specifically, Fig. 5.27 shows that while for

	International Standards	Regulations	Private standards
Governance and implementation	Set by private-public partnerships, including CSOs, academia, regulators, and industry representatives. Compliance with relevant requirements is in principle voluntary, but can become mandatory, e.g. when standards are incorporated in regulations.	Set, adopted, and applied by the government directly or indirectly. May be based on standards, particularly "international", but also "private" standards. Compliance with relevant requirements is mandatory.	Set and operated by private companies, CSOs, or joint initiatives thereof. Compliance with requirements is voluntary, but it can become *de facto* mandatory when the standard setter, e.g. a large distributor, has a particularly dominant position in a given product or geographic market.
Examples	ISO/ International Electrotechnical Commission (IEC) standards; ASTM International standards.	European Union "Seal Ban" (Regulation EC No. 1007/2009).	The Marine Stewardship Council (MSC) standard for sustainable seafood, which started in 1996 as a joint initiative of Unilever and the World Wildlife Fund (WWF), and later evolved to include other public interest and commercial stakeholders in its Stakeholder Council.

Fig. 5.26 Core characteristics of international standards, regulations, and private standards
Source: International Centre for Trade and Sustainable Development (ICTSD) (2017)

	No specific protocol	Private Standard	International Standard
Carbon Footprint	✓	✓	✓
Emissions reduction	✓	✓	✓
Insetting	✓	✓	✗
Offsetting	✓	✓	✗
Carbon neutrality	✓	✓	✓

Fig. 5.27 Voluntary standards for Carbon Neutrality process
Source: Authors' own elaboration

Carbon Footprint measurement and emission reduction (e.g. ISO 14064-1, ISO 14067, GHG protocol, PAS 2050) and Carbon neutrality (e.g. PAS 2060 and the Australian Climate active) there are available also international standards, for Insetting and Offsetting there are only available private standards (e.g. for insetting the Insetting Program Standard (IPS) developed by International Insetting Platform developed and for offsetting the standard provided by Gold Standard or VCS) (Fig. 5.27).

5.6.1 *The PAS 2060: 2014 Specification for the Demonstration of Carbon Neutrality*

The Publicly Available Specification PAS 2060 was first launched in 2010 by the British Standard Institution (BSI) and then revised in 2014. The PAS 2060 is the first carbon neutral certification that provide specifications to comply with for companies that wish to claim carbon neutrality (Thorn et al. 2011). Carbon Neutrality is defined by this standard as "*the condition in which during a specified period there has been no net increase in the global emission of GHGs to the atmosphere as a result of the GHG emissions associated with the subject during the same period*" (BSI 2010). This standard is applicable to activities, products, services, buildings, projects, towns, cities, and events (Birkenberg 2017) (Fig. 5.28).

Put into an equation, carbon neutrality can be expressed as illustrated in Fig. 5.29: the amount of GHG emissions calculated minus the emissions reduced through new technologies or processes and the amount of emissions offsetted (Fig. 5.29).

There are four steps involved in the certification process: measurement, reduction, offsetting, and validation and declaration (Fig. 5.30).

The first step is **the quantification of GHG emissions**, using an LCA based protocol to calculate the carbon footprint as for example the PAS 2050, the ISO 14064-1, or the GHG protocol. The carbon footprint measurements should include 100% of Scope 1 and Scope 2 emissions plus all Scope 3 emissions that contribute more than 1% of the total footprint, defined by the World Resources Institute and the

Fig. 5.28 PAS 2060: 2014 Specification for the demonstration of carbon neutrality
Source: BSI (2014)

Fig. 5.29 Carbon Neutrality Equation
Source: Carbon Clear (2011)

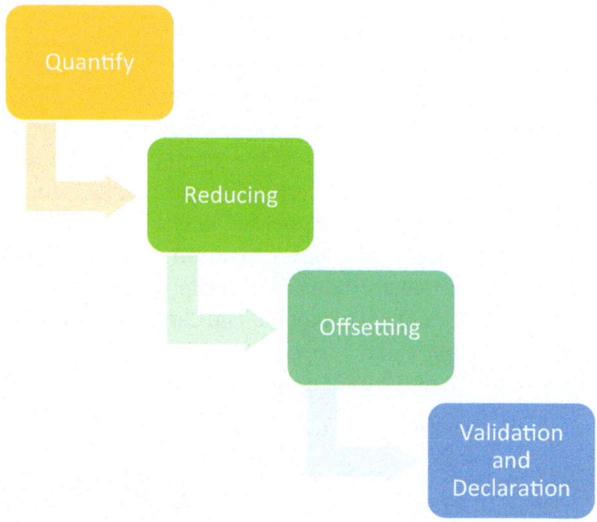

Fig. 5.30 The four steps towards CN
Source: Authors' own elaboration

World Business Council for Sustainable Development (2011) in the GHG protocol as:

- Scope 1- Direct GHG emissions – these occur from sources that are owned or controlled by the company, for example emissions from combustion in owned or controlled boilers, furnaces, vehicles, etc. or emissions from chemical production in owned or controlled process equipment.
- Scope 2- Electricity and heat indirect GHG emissions – this accounts for GHG emissions from the generation of purchased electricity and heat consumed by the company. Purchased electricity is defined as electricity that is purchased or otherwise brought into the organizational boundary of the company. Scope 2 emissions physically occur at the facility where the electricity is generated.
- Scope 3 - Other indirect GHG emissions – this is a reporting category that allows for the treatment of all other indirect emissions. Scope 3 emissions are a consequence of the activities of the company but occur from sources not owned or controlled by the company. Some examples of Scope 3 activities are the

extraction and production of purchased materials, the transportation of purchased fuels and the use of sold goods and services.

The second step include the development of a **Carbon Management Plan**, with a timescale for achieving carbon neutrality, a clear and accurate carbon reduction targets, and the offset strategy, this plan has to be presented and verified.

The third step is the **offsetting**. PAS 2060 requires to offset the remaining emissions, only through high-quality, independent third-party certified carbon credits (i.e. Clean Development Mechanism, Joint Implementation or Voluntary Carbon Standard etc.). Moreover, according to the PAS 2060 the carbon credits have to be genuinely additional and are retired after a maximum of 12 months to a credible registry.

The final step is the **validation and declaration of carbon neutrality.** The PAS 2060 prescribe the elaboration of a Qualifying Explanatory Statement (QES) in which the commitment to carbon neutrality and the achievement of carbon neutrality are declared. Additionally, the standard requires that all the documents that support carbon neutrality claims have to be publicly available. The PAS 2060 certification has a maximum validity period of 12 months, after which it can be renewed. The standard provides three different types of carbon neutral claim validation: self-validation, validation from another party, and independent validation by third parties (Carbon Clear 2011) (Fig. 5.31).

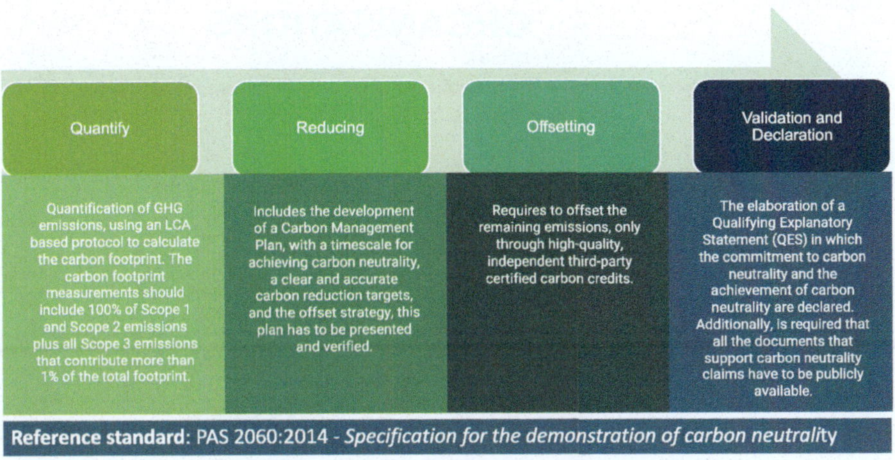

Fig. 5.31 Key steps for reaching and maintaining the carbon neutrality according to the reference standard
Source: Authors' own elaboration

5.6.2 The Climate Active Carbon Neutral Standard

Climate Active is an ongoing partnership between the Australian Government and Australian businesses to drive voluntary climate action. The Climate Active program and Climate Active Carbon Neutral Standard supports and guides businesses as they account for and reduce carbon emissions. **Climate Active Carbon Neutral Standard** (formerly the National Carbon Offset Standard, name has been changed in 2019) has been available to Australian businesses since 2010, with many of certified members claiming carbon neutrality since this date (Fig. 5.32).

Carbon neutral certification against the Climate Active certification is awarded to businesses and organizations that have credibly reached a state of achieving net zero emissions, otherwise known as carbon neutrality. This means that the activities associated with running a business or producing a particular product have no net negative impact on the climate. To achieve carbon neutral certification, entities must measure emissions, reduce these where possible, offset remaining emissions and publicly report on their carbon neutrality.

Fig. 5.32 Climate Active Carbon Neutral Standard for Organisations
Source: https://www.climateactive.org.au/be-climate-active

Fig. 5.33 Climate Active Carbon Neutral labels
Source: https://www.climateactive.org.au/be-climate-active

The climate active provide standards and certification that can be sought through the Australian Government's Department of the Environment and Energy for:

- Organisations
- Products & Services,
- Events
- Precincts

The carbon neutral certification is one of the most rigorous in the world and has been recognized by the European Union Commission and the World Bank as a mature and effective model to help businesses and incentivize emission reductions (Climate Active 2019). The Climate Active also provide labels to helps consumers identify and choose brands that are carbon neutral (Fig. 5.33).

Carbon neutral National standards can be consistent, integrated or independent from PAS 2060

- **Consistent:** The national standard has similar requirement of PAS 2060 but the certification according to PAS is not required;
- **Integrated:** The national standard has more requirements of PAS 2060 and the PAS certification is required;
- **Independent:** the national standard doesn't require PAS 2060.

The Climate Active Carbon Neutral Standard, for example, is consistent with the PAS 2060.

5.7 Verification, Certification, Accreditation and Labels

One way in which standards differ from each other is in their required means of verification and assessment of compliance. Standards can be compared on the governance and regulation of their scheme. This section explains how standards implement their requirements in terms of verification, certification and accreditation.

In Loconto and Dankers (2014) a graphical overview of how this aspect are related is provided (Fig. 5.34).

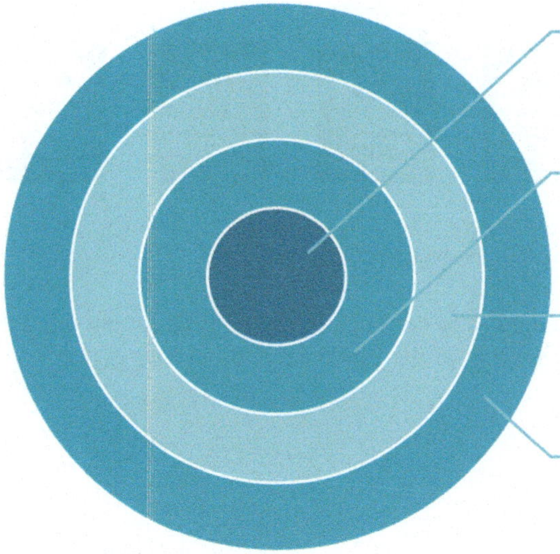

Fig. 5.34 Standards systems
Source: Loconto and Dankers (2014)

Verification

Verification is defined as "confirmation, through the provision of objective evidence, that specified requirements have been fulfilled. An assessor or inspector audits or verifies that requirements of a standard have been fulfilled. The results of verification are used as the basis for a decision on certification."(Vorley et al. 2010).

Audits can be of three types (Cook and Durivage 2018):

- **First-party audit** is performed within an organization to measure its strengths and weaknesses against its own procedures or methods and/or against external standards adopted by (voluntary) or imposed on (mandatory) the organization. A first-party audit is an internal audit conducted by auditors who are employed by the organization being audited but who have no vested interest in the audit results of the area being audited.
- A **second-party audit** is an external audit performed on a supplier by a customer or by a contracted organization on behalf of a customer. A contract is in place, and the goods or services are being, or will be, delivered. Second-party audits are subject to the rules of contract law, as they are providing contractual direction from the customer to the supplier. Second-party audits tend to be more formal than first-party audits because audit results could influence the customer's purchasing decisions.
- A **third-party audit** is performed by an audit organization independent of the customer-supplier relationship and is free of any conflict of interest. Independence of the audit organization is a key component of a third-party audit. Third-party audits may result in certification, registration, recognition, an award, license

approval, a citation, a fine, or a penalty issued by the third-party organization or an interested party

Certification

Certification is the procedure by which a certification body or certifier gives written or equivalent assurance that a product, process or service conforms to certain standards. There are three main types of certification:

There are three main types of certification (Washington and Ababouch 2011):

- **first-party certification:** by which a single company or stakeholder group develops its own standards, analyses its own performance, and reports on its compliance, which is therefore self-declared;
- **second-party certification:** where an industry or trade association or NGO develops standards. Compliance is verified through internal audit procedures or by engaging external certifiers to audit and report on compliance; and
- **third-party certification:** where an accredited external, independent, certification body, which is not involved in standards setting or has any other conflict of interest, analyses the performance of involved parties, and reports on compliance.

The organization performing the certification is called a **certification body** or **certifier**. The certification decision, i.e. the granting of the written assurance or "certificate", is based on the inspection report, possibly complemented by other information sources. The granting of a certificate of conformity may be subcontracted by the standard owner to the certification body or issued by the standard owner itself.

Accreditation

To ensure that the certification bodies have the capacity to carry out certification programs, they are evaluated and accredited by an authoritative body. Certification bodies may have to be accredited by a governmental or parastatal institute, which evaluates compliance with guidelines set by ISO, the European Union or some other entity for the operation of certification and inspection bodies. In addition, standard-setting bodies might accredit certification bodies for the scope of their particular standard. When the standard-setting body has developed normative standards, they will evaluate whether the specific standard used by the certification body is in line with the generic standard and whether they are satisfied with the method of verification.

Labels

Certification label is a label or symbol indicating that compliance with standards has been verified. Use of the label is usually controlled by the standard-setting body. Where certification bodies certify against their own specific standards, the label can be owned by the certification body. While the certificate is a form of communication between seller and buyer, the label is a form of communication with the end consumer. Labels can be owned by the company, by the private standard or by the international standard.

Figure 5.35 provide an overview of the different certification, verification and accreditation systems depending on the standard applied by the company.

To better explain how to governance of the carbon neutrality works nowadays we elaborated the case study of Gran Cereale.

Table 5.3 presents the main characteristics of the Gran Cereale Carbon neutrality strategy. Moreover, in Fig. 5.36 the main standard and the verification system used in the Gran Cereale Carbon neutrality strategy have been presented. As we can see from the Figure Gran Cereale used the Environmental Product Declaration (EPD) to

Methodology applicable	Certification			Verification			Accreditation of the CB	Label			Robustness
	1st party	2nd party	3th party	1st party	2nd party	3th party		Company Label	Private Label	International Label standard	
International Standard	✓	✓	✓	✓	✓	✓	✓	✓	✓	✓	★★★
Private Standard	✓	✓	✗	✓	✓	✗	✗	✓	✓	✗	★★
No specific protocol	✓	✗	✗	✓	✗	✗	✗	✓	✗	✗	★

Fig. 5.35 Overview of standards and certification systems
Source: Authors' own elaboration

Table 5.3 Gran Cereale Case study

Company	Gran Cerale
Scope of CN	Scope 1 +2 + 3
Year project started	2015
Carbon Footprint Measurement	Environmental Product Declaration (EPD) based on LCA methodology; 3rd party verified by Bureau Veritas
CN standard	PAS 2060; 3rd party verified by DNV
Reduction strategy	**Reduction:** Energy efficiency projects verified by the PAS auditor.
Offset strategy and certification	Strategy developped with AzzeroCO2, an Italian consulting company recognized in carbon neutral strategies, Gran Cereale has chosen a REED+ project labeled by the Verified Carbon Standard (VCS) verified by TUV. The compensation project selected by the Gran Cereale is to protect the forest in Marajo, Brazil. The project aims to reduce the deforestation of the island of Marajo, preserving 2,500 hectares and supporting the development of a local school and income-generating activities from the fruit trade. The project develops and promotes the abandonment of wood exploitation and transport by river, with the consequent failure to issue 72,388 tons of CO2 equivalent emissions per year
Label	no

Source: Authors' own elaboration

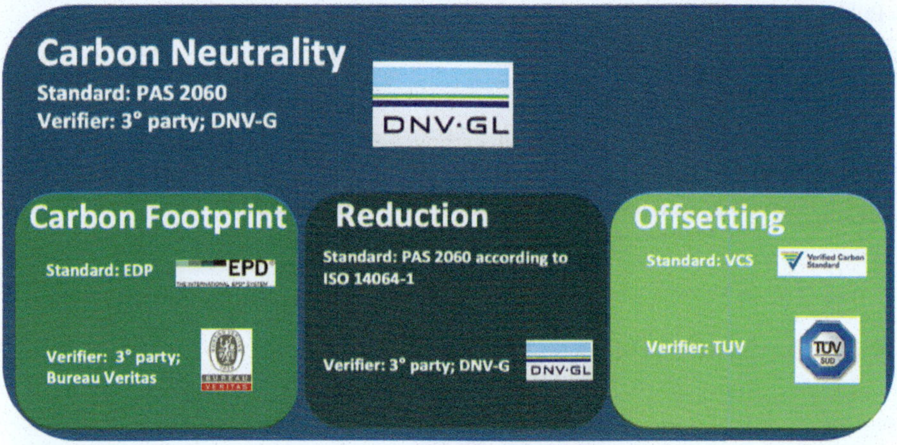

Fig. 5.36 Gran Cereale Standards and verification system
Source: Authors' own elaboration

assess its carbon footprint. The EPD has been third-party verified and certified by accredited control body Bureau Veritas. Then, the emission reduction plan that has been elaborated following the PAS 2060 requirements in accordance with the ISO 14064-1 has been verified during the third-party PAS 2060 audit by the accredited control body DNV-GL. Moreover, the offsetting strategy has been carried out by buying carbon credits that are certified trough the private standard VCS and third-party verified by the accredited control body TUV. Finally, the state of Carbon neutrality has been third-party certified according to the international standard PAS 2060 and third-party verified by the accredited control body DNV-GL.

5.8 Carbon Neutral Labels

Eco-labels schemes can be characterized according to the ownership of the standard, notably whether they are private, public, non-profit or hybrid. They can be developed by private entities, by public agencies, or jointly by stakeholders and experts from the public and private sectors.

Below some examples of carbon neutrality private driven and public driven labels.

Private Driven

- Certification Bodies
- Companies
- Consultant

Public Driven

- International Organizations
- Governments

We elaborated two examples of carbon neutral labels, using the information contained in the Carbon neutrality framework, one of a public driven carbon neutral standard and one of a private driven carbon neutral standard.

Public Driven Standard

- **Provider Name**: Australian Government; **Location**: Australia**; Number of clients**: 100; **Number of Agri-food Companies Served**: 5; **Geographical area of Influence**: Australia; **Label Name**: Climate Active (Ex NCOS); **Year of creation**: 2010; **Scope**: Organizations, products & services, events and precincts; **Verification of the carbon footprint**: yes; **Verification of the offset program**: yes; **CF Methodology**: GHG Protocol, ISO 14064, ISO 14040, ISO 14044; **Accreditation system of the CB**: yes; **Control system:** third party required for full neutrality process**; Main control bodies**: "Included on the Register of Greenhouse and Energy Auditors OR accredited to the international standard ISO 14065:2013 OR accredited to recognized international standards based on ISO 14040."

Private Driven Standard

- **Provider Name**: Carbon Trust; **Location**: UK**; Geographical area of Influence**: World; **Carbon footprint analysis:** yes**; Set targets and reduce emissions:** yes**; Offsetting:** yes**; Insetting:** no**; Label Name**: Carbon Trust Carbon Neutral; **Scope**: Company, sector, product; **Verification of the carbon footprint**: yes; **Verification of the offset program**: yes; **The company can invest in chosen sustainable compensation projects**: no; **Certification of generated/purchased carbon credits is compulsory**: yes; **CF Methodology**: PAS 2060; **Offset certification standard**: GS; **Accreditation system of the CB**: yes; **Control system:** Verification by the label owners

Moreover, Table 5.4 presents the main characteristics the main carbon neutral labels available for companies that want to certify their commitment towards carbon neutrality goal.

Table 5.4 Carbon labels

Label	Owner	Logo	Objectives	Scope	Control system
Carbon Neutral® certification	Natural Capital Partners		Certify zero net emissions	Company, sector, product, event	3rd party required for all neutrality process
Carbon Trust Carbon Neutral Certification	Carbon Trust		Certify zero net emissions	Company, sector, product	Verification by the label owner
NoCO2 certification	Carbon Reduction Institute		Certify zero net emissions	Company, sector, product	Verification by the label owner
CO2 Neutral Label	CO2Logic		Certify zero net emissions	Company	3rd party required for all neutrality process
Climate Neutral Label	Fondation MyClimate		Certify zero net emissions	Company, sector, product, event	Verification by the label owner
Climate Neutral Now	UN Climate Change		Certify zero net emissions	Company, organizations, governments, citizens	3rd party required for all neutrality process

(continued)

Table 5.4 (continued)

Label	Owner	Logo	Objectives	Scope	Control system
Certified Carbon Neutral	SCS Global Services		Certify zero net emissions	Organization, product or brand	Verification by the label owner
Carbon Progress	Bureau Veritas		Reduce carbon emissions	Companies, public organizations	Not specified
Engagement Climat	ECOCERT		Reduce carbon emissions	Companies, public organizations	Verification by the label owner
Carbono neutro certificado Huella de Carbono	Icontec		Certify zero net emissions (Carbono neutro); Reduce carbon emissions (huella de carbono)	Companies, product, services; process	Verification by the label owner
AENOR Medio Ambiente de Emisiones de CO2	AENOR		Reduce carbon emissions; Compensated carbon emissions; calculated carbon emissions	Products & services	3rd party required for all neutrality process
NCOS	Australian Government		Certify zero net emissions	Organizations, products & services, events and precincts	3rd party required for all neutrality process

Net zero initiative	Carbon 4, Net zero initiative		Certify zero net emissions	Organizations	Not specified
Climate Neutral	South Pole		Certify zero net emissions	Companies, products, events	3rd party required for all neutrality process
Klimaatneutraal product	Climate Neutral Group		Certify zero net emissions	Organizations	Verification by the label owner
CarbonNzero and Cemars	Enviro-Mark Solution		Carbon reduction (Cemars) and carbon neutrality (CarbonNzero)	Companies	Verification by the label owner
VOCAL	French Government		Carbon reduction (Cemars) and carbon neutrality (CarbonNzero)	Public or Private organizations	3rd party required for all neutrality process
CO2 Neutral	NEPCon		Certify zero net emissions	Corporate or products	3rd party required for all neutrality process
Carbon Neutral	Carbon Footprint Ltd.		Certify zero net emissions	Organization, product, service, event	3rd party required for all neutrality process

(continued)

Table 5.4 (continued)

Label	Owner	Logo	Objectives	Scope	Control system
		carbon footprint standard Carbon Neutral PAS 2060			
Carbon neutral quality mark	Quality Assurance Standard	QAS CARBON NEUTRAL 2015	Certify zero net emissions	Organization	3rd party required for all neutrality process
Plan Vivo	Plan Vivo Foundation	Plan Vivo	Certify reduction or avoidance of one metric tons of carbon dioxide, plus livelihood and ecosystem benefits	Projects	3rd party required for all neutrality process
Certified Carbon Neutral	Verus Carbon Neutral	FLEET CERTIFIED CARBON NEUTRAL	Certify zero net emissions	Organization, product, sites, transportation, Group of business	Verification by the label owner
Climate neutral	Climatepartner	Climate neutral Product	Certify zero net emissions	Companies, products, packaging, website, printing	3rd party required for all neutrality process

References

Birkenberg AS (2017) The potential of certification for climate change mitigation in the agri-food sector – a case study of carbon neutral certified coffee from Costa Rica Dissertation 183

BSI (2010) PAS 2060:2010 Specification for the demonstration of carbon neutrality

BSI (2014) PAS 2060:2014 Specification for the demonstration of carbon neutrality

Carbon Clear (2011) White paper: PAS 2060, the first standard for carbon neutrality

Carbon Market Watch (2019) Carbon markets 101: the ultimate guide to global offsetting mechanisms, pp 1–12

Climate Active (2019) Be Climate Active

Climate Smart Group (2017) Assessing the insetting landscape: the current status of insetting in the sustainability industry

Cook E, Durivage MA (2018) The ASQ CSQP study guide

Davies M (2016) Insetting: developing carbon offset projects within a company's own supply chain and supply chain communities [WWW document]. ICROA University, Bristol

Forest Trends' Ecosystem Marketplace (2019) Financing emissions reductions for the future: state of the voluntary carbon markets 2019

International Centre for Trade and Sustainable Development (ICTSD) (2017) Private standards, trade, and sustainable development: policy options for collective action

International Emissions Trading Association (2019) Offsets: the basics

International Insetting Platform (2017) Insetting program standard (IPS), pp 1–37

International Organization for Standardization/International Electrotechnical Commission (ISO/IEC) (1996) Guide 2: Standardization and related activities – General vocabulary

International Organization for Standardization/International Electrotechnical Commission (ISO/IEC) (2001) Directives, Part 2: principles and rules for the structure and drafting of ISO and IEC documents

International Trade Center (2019) WTO: non-tariff measures – technical barriers to trade Part 3: Di

Loconto A, Dankers C (2014) Impact of voluntary standards on smallholder market participation in developing countries. Food And Agriculture Organization Of The United Nations

Phelan LT (2015) Adding value to smallholder cattle value chains in Nicaragua, in the context of carbon insetting, pp 1–171

Plan Vivo (2019) Insetting: a user guide

Rousset S, Deconinck K, Jeong H, Von Lampe M (2015) Voluntary environmental and organic standards in agriculture: policy implications. OECD 86

Thorn MJ, Kraus JL, Parker DR (2011) Life-cycle assessment strengths, weaknesses, management tool: as a sustainability and other considerations. Environ Qual Manag. https://doi.org/10.1002/tqem

Tipper R, Coad N, Burnett J (2009) Is "insetting" the new "offsetting"? Econometrica

Vorley, B., Beekmans, A., Hormer, S., 2010. Food-related voluntary sustainability standards: a strategy guide for policy makers

Washington S, Ababouch L (2011) Private standards and certification in fisheries and aquaculture: current practice and emerging issues. FAO Fisheries and Aquaculture Technical Paper. https://doi.org/10.1017/CBO9781107415324.004

World Resources Institute & World Business Council for Sustainable Development (2011) Greenhouse gas protocol. product life cycle accounting and reporting standard. World Resour. Institute, Washington, DC

Chapter 6
How Difficult Is It to Reach Carbon Neutrality? Tales of Different Agri-Food Chains

The sixth chapter presents the different actors of the agri-food chain involved in the adoption of carbon neutral/low carbon approaches. Then, the contribution of these actors to climate change and agri-food sector emission has been analysed. Additionally, the chapter covers analysis from the literature on different food chains and how much they emit at different stages of the value chain. Moreover, the chapter presents an analysis of the main actions that the various actors of the supply chain can take to make their activity climate-friendly and therefore reduce the emissions associated with their main assets. Finally, three company case studies have been presented to discuss differences across value chains (fruits, coffee, and meat).

6.1 The Role of Different Actors of the Agri-Food Supply Chain in Achieving Sector' Carbon Neutrality

To fight climate change and mitigating its impacts on agri-food sector is necessary to consider all the supply chain (BSR 2009). Carbon emissions associated with agri-food sector arise along the supply chain from agricultural production to the processing, wholesale and retail levels, to the consumer and finally at the waste management level. The agri-food supply is mainly composed by input companies, farmers, traders, food companies and retailers and consumers (Fig. 6.1).

These different actors contribute in a diversified way to GHG emissions in the sector. However, the cooperation between these actors and the role that each of them play in the transition to the low carbon economy is critical for understanding and identifying diversified mitigation strategies. Analysis for determining agri-food sector carbon impacts are very difficult and characterized by high uncertainty mostly due to increased variability in management practices, the case-dependent conditions, different types of food products, and different calculation methods. In this sense, Noleppa (2012) conducted a meta-analysis synthetizing different studies' results

A. Acampora et al., *Carbon Neutrality in the Agri-food Sector*,
https://doi.org/10.1007/978-3-030-88048-4_6

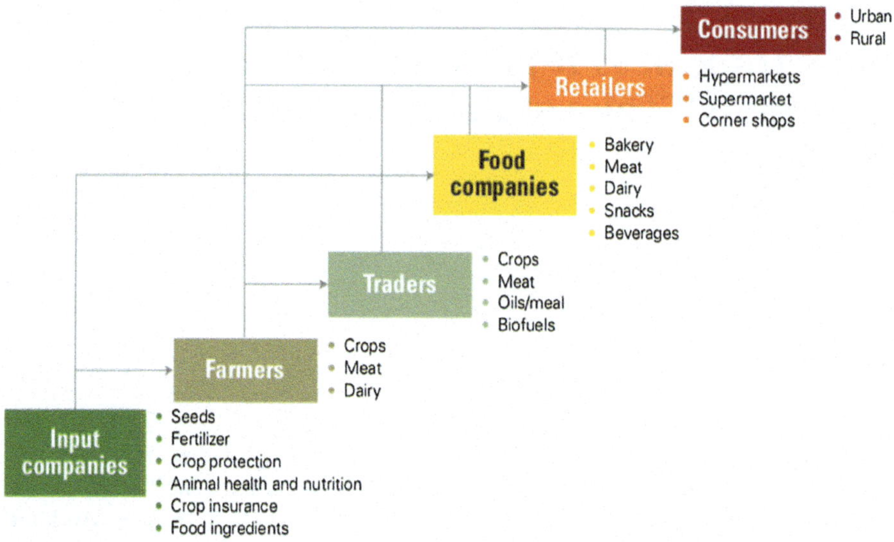

Fig. 6.1 Agri-food supply chain actors
Source: KPMG (2013)

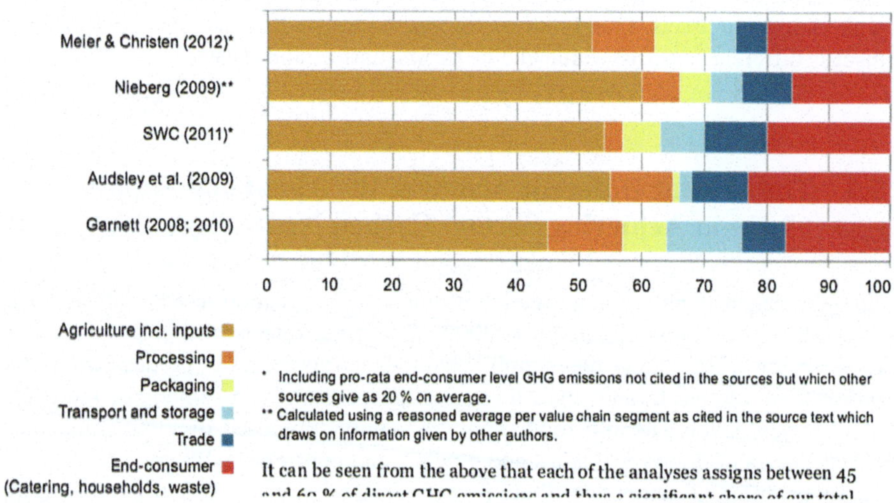

Fig. 6.2 Shares of food supply chain stages in food-related GHG emissions (excluded land use change)
Source: Noleppa (2012)

accounting for GHG emission of the food supply chain (Audsley et al. 2009; Barners-Lee 2012; Garnett 2008; Meier and Christen 2012; Nieberg 2009). Figure 6.2 shows Noleppa (2012) results in relation to the contribution in terms of GHG emission of each food supply chain stages.

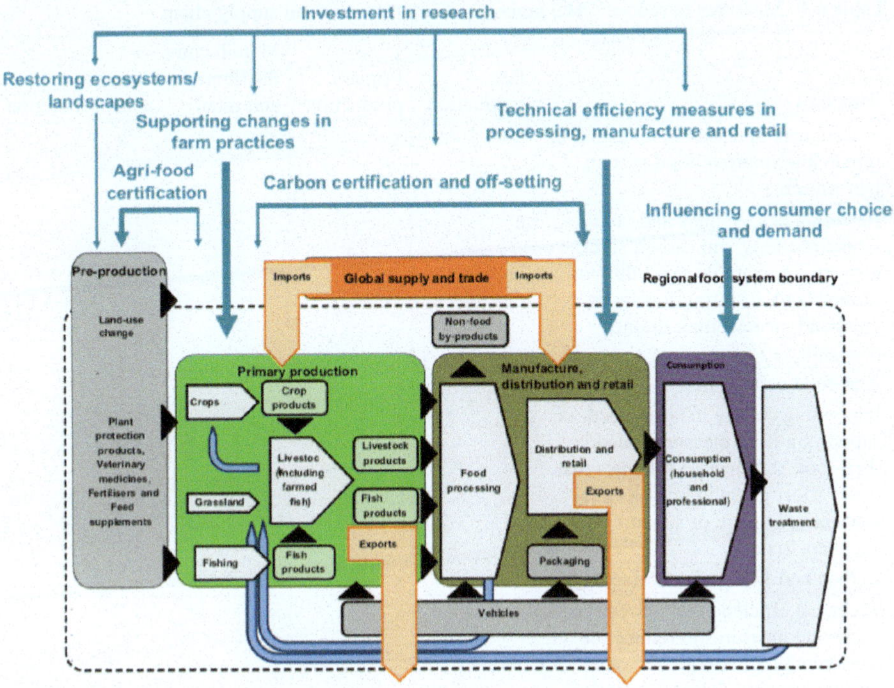

Fig. 6.3 CSR practices and agri-food supply chain
Source: Murphy-bokern and Kleemann (2014)

Considering these aspects, Murphy-bokern and Kleemann (2014) explored the impact of different CSR practices that can be implemented by the actors of the chain (Fig. 6.3). In Fig. 6.3 the arrows in blue represent the greenhouse gas mitigation impact pathways in relation to the different actors and stage of the agri-food supply chain.

The authors highlighted the following measures to reduce GHG emissions in different supply chain stages and to support a low carbon strategy summarized in Table 6.1.

Subsequently, we have summarized for literature the main actions that the various actors of the supply chain can take to make their activity climate friendly and therefore reduce the emissions associated with their main assets.

- **Input companies**

  Input companies can reduce their carbon load through:

- Boosting research on stress and heat tolerant crops (KPMG 2013)
- Produce more resilient seeds and varieties (FAO 2018; Niles et al. 2017)
- Increase nitrogen use efficiency in fertilizer production (KPMG 2013)
- Increase the efficiency of input production systems, shift toward renewable or less GHG-intensive energy sources for its production, or shift toward more

Table 6.1 Measures to reduce GHG emission along the agri-food supply chain

Measures	Pre-production	Primary production	Manufacture, distribution and retail	Consumption
Investement in research: Investing in research to improve food system performace.	x	x	x	x
Restoring eco-systems and land-scape: restoring soil carbon stocks through for example re-wilding of farmed land, restoring forests or grassland, or rewetting drained peatland.	x			
Agri-food certification	x	x		
Supporting change in farm practices: support on-farm measures outside Agri-food certification schemes that could affect greenhouse gas emissions or pressurise or support their suppliers to do so		x		
Carbon certification and offsetting		x	x	
Technical efficiency measures: Technical measures to reduce the consumption of energy and other resources within the post-farm supply chain		x	x	
Influencing consumer choice and demand				x

Source: Authors' own elaboration on Murphy-bokern and Kleemann (2014)

sustainable agricultural inputs required for agroecological practices (which would occur in the production phase) (Niles et al. 2017, 2018).

- **Farmers/Production**

Farming activities are the most impactful in terms of GHG emission along the supply chain. Therefore, farmers play a pivotal role in the law carbon transition. The main actions that farmers can undertake are listed below:

- Reducing deforestation and boost reforestation and enhancing carbon removals (Muller et al. 2011; Smith et al. 2014)
- Optimizing nutrient use: precise dosage and timing when applying organic and inorganic fertilizers; incorporating nitrogen-fixing legumes into rotations (Garnett 2011)
- Improving productivity: approaches that increase the yield of edible output per unit of emissions generated including: crop and animal breeding; feed optimization and dietary additives; pest and disease management(Garnett 2011)
- Managing and benefiting from the outputs: including manure and plant biomass: composting, and the use of anaerobic digestion (Garnett 2011)

- Reducing the carbon intensity of fuel inputs through energy efficiency improvements and the use of alternative fuels such as biomass, biogas, wind and solar power(Garnett 2011)
- Reduction of food wastage and storage losses (Muller et al. 2011; Smith et al. 2014)
- Switch from ruminants to monogastric animals (Pigs, poultry)(Muller et al. 2011)
- Use of energy efficient machinery and optimized machinery use (Muller et al. 2011)
- Switch to organic farming (FAO 2011)
- Eco-labelling, carbon labelling and carbon offset (Murphy-bokern and Kleemann 2014)

- **Traders**

  Agricultural trading companies could help to mitigate climate change by:

- Allowing to move agri-food production to most efficient regions in terms of GHG emissions (Zimmermann et al. 2018);
- Using dedicated (most suitable) regions for the production of (new) energy crops (Zimmermann et al. 2018);
- Trade could help mitigate climate change by fostering the exchange and distribution of climate friendly technology and solutions (Zimmermann et al. 2018);
- Traders may foster the diffusion of carbon labels certifications boosting collaboration between farmers and food processor and by promoting carbon labels sustainable procurements (Murphy-bokern and Kleemann 2014)

- **Food companies**

  Food manufacturing and processing companies can reduce their associated GHG emissions through:

- Apply energy efficiency measures (Fischedick et al. 2014)
- lower-emission modes of transport (Garnett 2011)
- Increasing and contributing to renewable energy procurement and bio-energy (FoodDrinkEurope 2015)
- Reducing emissions from transport and logistics (Fischedick et al. 2014; FoodDrinkEurope 2015)
- Improve efficiency of food storage (FAO 2011)
- Minimizing packaging of food (FAO 2011)
- Resource efficiency: reducing unnecessary use of products and equipment; recycling and reuse where environmentally appropriate (Garnett 2011)
- Eco-labelling, carbon labelling and carbon offset (Murphy-bokern and Kleemann 2014)
- Set up environmental and carbon requirements for products and suppliers (Styles et al. 2012)
- Help consumers to make informed choices and engage them in choosing carbon labeled products (Elofsson et al. 2016; Leire and Thidell 2005; Sharp and Wheeler 2013)

- **Retailers**

The role of retailers in supporting low carbon transition and carbon labels diffusion from a consumer perspective will be analyzed in Sect. 8. However, in this paragraph we will synthetize the main actions that can be implemented by retailers:

- Raise consumer awareness on the point of sale (Retail Forum for Sustainability 2011; Sharp and Wheeler 2013)
- Reduce food waste (Deloitte 2016)
- Set up internal policies to measure and reduce the direct and indirect emissions deriving from their operations in stores (Retail Forum for Sustainability 2011)
- Promote frontrunner eco-labelled, carbon labelled and low carbon products (Styles et al. 2012)
- Set up environmental and carbon requirements for products and suppliers (Styles et al. 2012)
- Choice editing and sustainable and green procurement. Exclude worst performing products, and require widespread certification according to environmental standards for priority product groups. (Styles et al. 2012)
- Strategic collaboration on product and standard development(Styles et al. 2012)
- Provide consumers with carbon information at point of sale (POS)(Elofsson et al. 2016; Leire and Thidell 2005; Sharp and Wheeler 2013)
- Help consumers to make informed choices and engage them in choosing carbon labeled products (Elofsson et al. 2016; Leire and Thidell 2005; Sharp and Wheeler 2013)

- **Consumers**

As we have also argued in the previous section, the role of consumers in the fight against climate change is fundamental. The main actions that consumers can undertake are listed below:

- Dietary change: vegan, vegetarian, eat less meat and dairy products (Climate-Smart Agriculture Leadership Forum 2016; Poore and Nemecek 2018)
- Avoid high-impact producers and products (Poore and Nemecek 2018)
- Prefer carbon labelled or eco-labelled products (Hartikainen et al. 2014; Vanclay et al. 2011)
- Avoid overconsumption (Gladek et al. 2018)
- Cook and store foods in energy conserving ways (Garnett 2011)
- Reduce food waste (Austgulen et al. 2018; Garnett 2011).

6.2 Differentiated Impacts for Different Food Products

As we explored also in Sect. 2.1 food products have different environmental impacts. This is particularly true for their Carbon footprint (CF). The CF of different food products is highly variable, even for the same food product, depending on the production system and methodological choices in the CF assessment. However, one pattern that has emerged is that livestock products generally have a considerably larger CF than plant-based foods (European Commission 2006), although high CF values for plant-based foods have been identified for some products that are produced in heated greenhouses, transported by air or produced in low-yielding systems (Stoessel et al. 2012). Beef and lamb meat have exceptionally high CF, followed by cheese, due to the contribution of CH_4 from enteric fermentation in ruminants. Meat from monogastric animals, such as pigs and poultry, shows lower CF values than products from ruminants, but still higher than most foods of vegetal origin, due the large amount of feed needed in livestock production and emissions from manure handling. These numbers do not include emissions from LUC or from carbon sequestration in soils. Another important insight from studying the CF of food is that direct emissions from livestock production are dominated by pre-farm and on-farm emissions, while post-farm emissions are often considerably smaller (Cederberg et al. 2009; Peters et al. 2010). For plant-based foods, post-farm stages can make a significant contribution to the total CF, e.g. emissions from transport can be a major contributor to the total CF for fruit and vegetables (Sim et al. 2007; Weber and Matthews 2008). In this Section to analyze these differences we used as reference point the Double Pyramids databases elaborated by the Barilla Center for Food & Nutrition (Fig. 6.4). Proposed by the Barilla Center for Food & Nutrition in June 2009 and now in its seventh edition, the Double Pyramid elaborated a model of sustainable food choices consistent with both health and environmental criteria. These pyramids allow the comparison between the classic Food Pyramid (constructed on the basis of nutritional properties of foods) and the Environmental Pyramid (in which each food is positioned according to its environmental impact).

The key message behind the Double Pyramid is the relationship between the environmental impact of food production and consumption, and its nutritional aspects.

The food section of the Double Pyramid was derived by pooling different international nutritional guidelines which can be traced back to the model known as the "Mediterranean Diet" that stands out for its completeness and nutritional balance. The food pyramid narrows towards the top, recommending less frequent consumption of foods with increasing energy density, usually made from animal proteins, fats, and simple sugars.

The environmental section of the Double Pyramid, instead, reclassify food not in accordance with its nutritional characteristics but in relation to its impact on the environment: using data on impact per kilo of product yields. According to these data the pyramid has been inverted, with foods which have a greater environmental impact at the top and those with a reduced impact at the bottom. By combining the

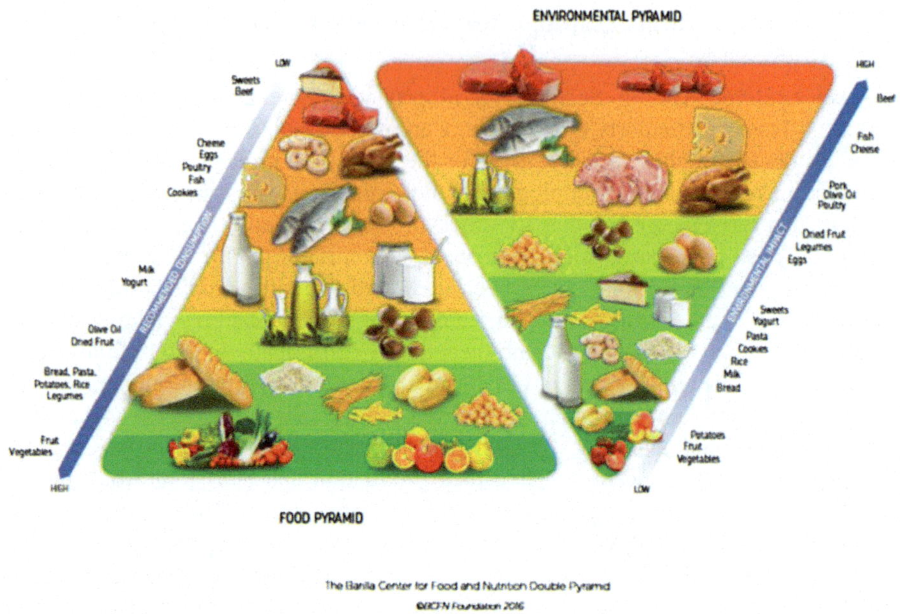

Fig. 6.4 BCFN Double Pyramids
Source: Barilla Center for Food and Nutrition (2016)

two pyramids, a single dietary pattern results which clearly matches two different but equally important objectives: health and environmental protection.

In the technical document of the BCFN that is behind the Double Pyramids construction three different pyramids have been elaborated: The Water Footprint, the Carbon Footprint and the Ecological Footprint.

- The **Carbon Footprint**, which identifies the greenhouse gas emissions responsible for climate change, and is measured as mass of carbon dioxide equivalent [CO2eq];
- The **Water Footprint**, which quantifies water consumption and how the resource is utilized, and is measured in volume, liters or cubic meters [m3], of water;
- The **Ecological Footprint**, which calculates the area of biologically productive land required to provide the resources and absorb the emissions associated with a production system, and is measured in square meters [m2] or global hectares [global ha].

However, the Environmental Pyramid is constructed using the Ecological Footprint alone, selected for its ease of communication associated with the unit of measure used [m2].

These pyramids are build calculating the environmental impact associated with each food is assessed using data calculated by means of the **Life Cycle Assessment (LCA) analysis**.

SOURCE TYPE	REFERENCE	RELIABILITY
LCA database	Ecolnvent	Public information, used by professionals. Quality can be variable; generally, the information is not specific to a manufacturer but generalized to the product at hand.
	LCA food	
	Water Footprint Network	
	Ecological Footprint Network	
Verified publication	EPD™ Climate Declaration™	Information validated by third party. It may be specific to a single manufacturer.
Scientific publication - executive summary - working paper - presentation	Complete list in bibliography	Information on a scientific work validated by a head reviewer. Product-specific but generally reliable in terms of quality.
Internal elaboration	-	Processing executed specifically for this work. Since it is minimized as far as possible and uses only public data, it proves less reliable than the other sources cited.

Fig. 6.5 Sources used to build the pyramids
Source: Barilla Center for Food and Nutrition (2016)

This takes the entire production chain into account and studies all its phases: production of the raw materials, manufacturing, packaging, transportation, distribution, and consumption of food. The Environmental Pyramid is based entirely on public data and information from scientific literature, producers' publications, and databases typically consulted in life cycle analysis. After an initial consistency check with the transparency requirements imposed by the BCFN procedure, the data was subjected to a "logical reorganization" in order to construct the schematic diagram (Fig. 6.5).

Over the years, the amount of available data has greatly increased, making the scientific basis much more reliable and resilient. This seventh edition features over 1300 sources of data, as opposed to the 140 cited in the first edition (Fig. 6.6). For each of the three environmental indicators, the percentile distribution of each macro category which makes up the environmental pyramids is indicated (Fig. 6.7).

In the three environmental pyramids, the values were used to determine the "classification" of the impacts of food while the specified impact intervals for the construction of the pyramids were partially modified for a more even distribution of food among the groups. For the three pyramids emerge clearly that meat and cheese are characterized by significant impacts, while fruit and vegetables have lower environmental impact values (Figs. 6.8 and 6.9).

For what concerns this study we will analyze more in-depth the Carbon Footprint Pyramids as it analyzes the GHG emissions associated with different food products. Figure 6.9 shows in darker orange the emissions associated with the cooking phase. Then it presents with red bars the range of emissions associated with different food found in different academic studies from the minimum values when the red bars start to the maximum values when the red bars finish. Moreover, the black bar represents

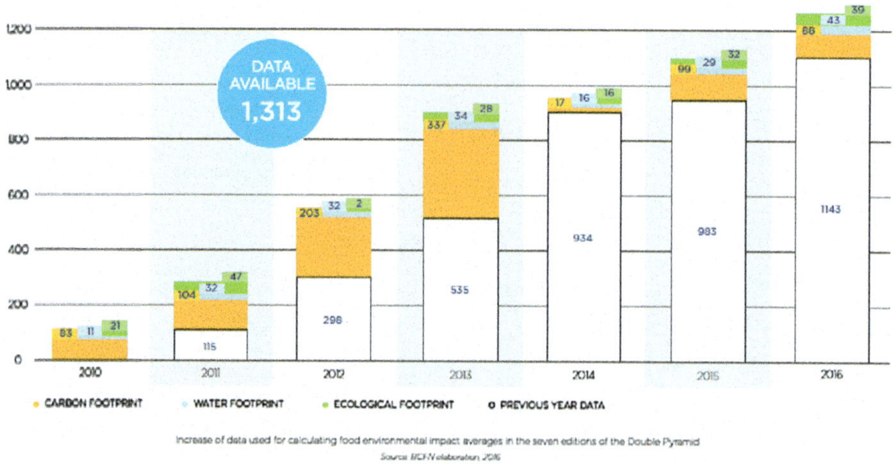

Fig. 6.6 Number of sources used to build the pyramids database (by years and for different types of footprint)
Source: Barilla Center for Food and Nutrition (2016)

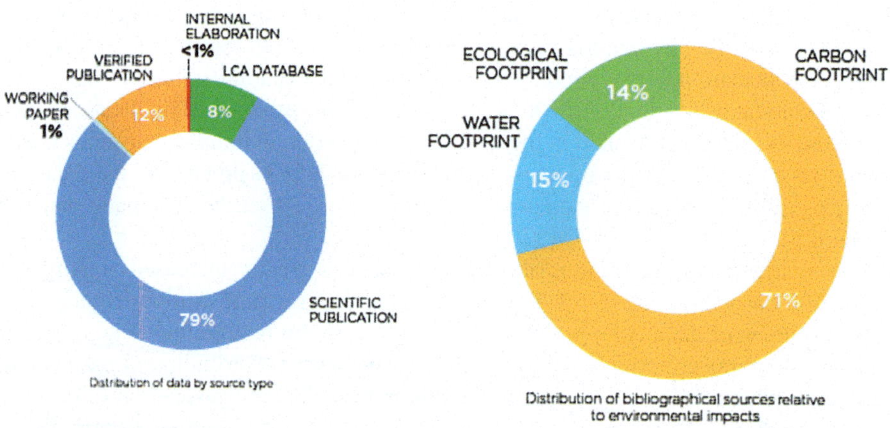

Fig. 6.7 Distribution of data by source type and distribution of bibliographical sources relative to environmental impacts
Source: Barilla Center for Food and Nutrition (2016)

the mean value. Finally, at the end of each food bar, the average value of GHG emissions associated with different food products has been reported.

As we can see from Fig. 6.9 beef (25.895 g CO2 eq/Kg) has the highest carbon impact with cheese (9.250 g CO2 eq/Kg) and butter (8.305 g CO2 eq/Kg). Instead, fruit (490 g CO2 eq/Kg), sugar (600 g CO2 eq/Kg) and vegetables (775 g CO2 eq/Kg) have the lowest impacts (Fig. 6.9) (Barilla Center for Food and Nutrition, 2016).

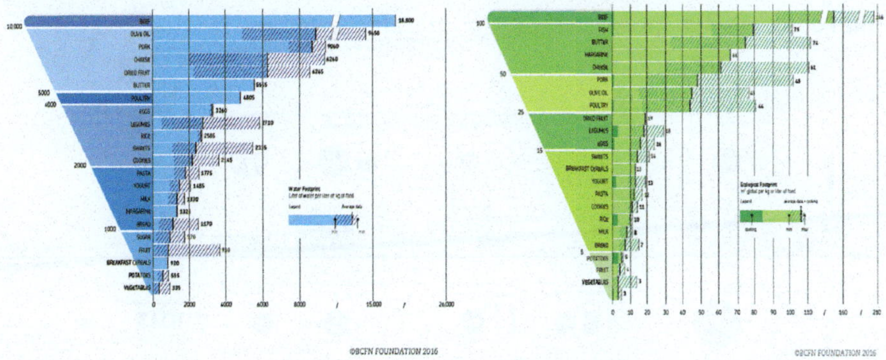

Fig. 6.8 The Water Footprint and Environmental Footprint pyramids
Sourc: Barilla Center for Food and Nutrition (2016)

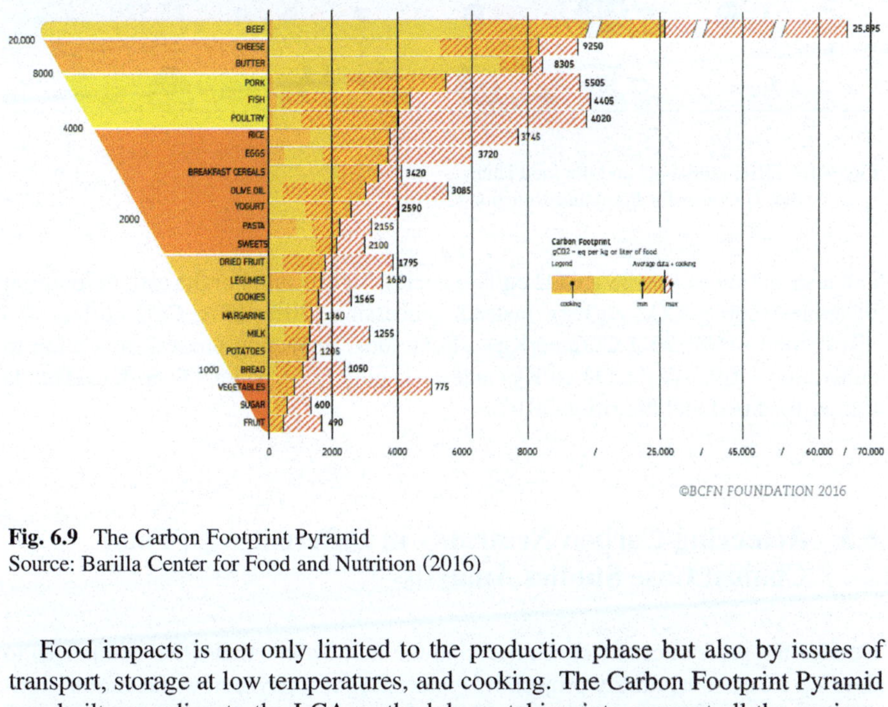

Fig. 6.9 The Carbon Footprint Pyramid
Source: Barilla Center for Food and Nutrition (2016)

Food impacts is not only limited to the production phase but also by issues of transport, storage at low temperatures, and cooking. The Carbon Footprint Pyramid was built according to the LCA methodology, taking into account all the environmental impacts generated by different food products throughout their entire life cycle, from farm to fork. Figure 6.10 analyze for red meat, pasta and apple each phase of the food chain and their contribution to the Carbon Footprint. From the figure emerges as for the red meat carbon footprint (23.328 g CO2 eq/Kg) the majority of the emissions are concentrated in the livestock farm (72%; 16.900 g CO2 eq/Kg) and pasture (17%; 3.900 g CO2 eq/Kg), and in proportion the emissions related to packaging (1%; 200 g CO2 eq/Kg), processing (3%; 700 g CO2 eq/Kg)

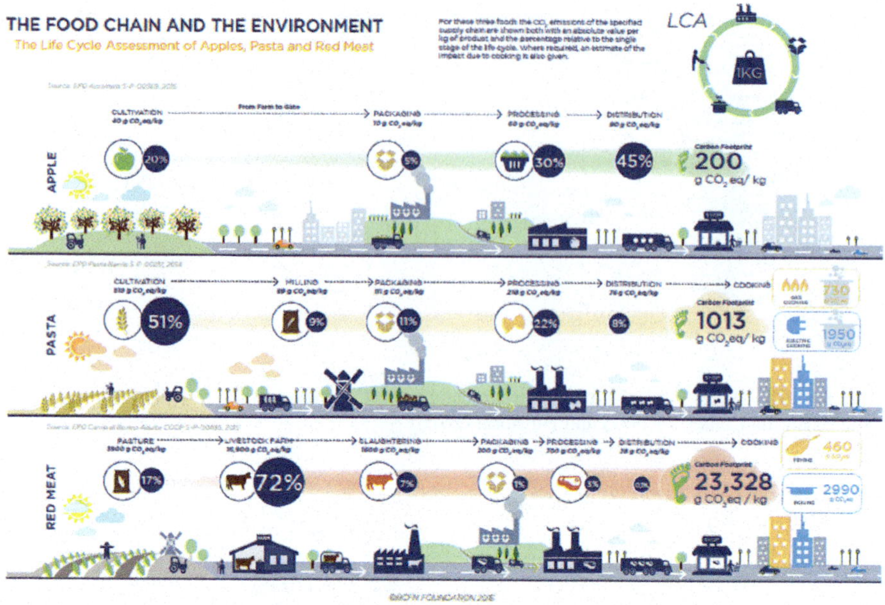

Fig. 6.10 GHG emissions and the food chain
Source: Barilla Center for Food and Nutrition (2016)

and distribution (0,1%; 28 g CO2 eq/Kg) are less significant. For the carbon footprint of apples (200 g CO2 eq/Kg), instead, processing (30%; 60 g CO2 eq/Kg) and distribution (45%; 90 g CO2 eq/Kg) are the most impacting phases, in relation to cultivation (20%; 40 g CO2 eq/Kg) and packaging (5%; 10 g CO2 eq/Kg) (Barilla Center for Food and Nutrition 2016).

6.3 Achieving Carbon Neutrality in Different Agri-Food Chains: Case Studies Analysis

Besides the preferences of consumers and the motivations that drive companies to develop a carbon neutrality project, it is interesting to analyze how the various agri-food businesses are approaching this tool. Unfortunately, being in its early phase of diffusion, case studies on carbon neutrality in the agri-food sector are scarce in literature.

However, some research papers exploring the application of carbon neutral approach in the agri-food sector have been presented for different types of products: meat (e.g. Mayberry et al. 2019), fruits (e.g. Kilian et al. 2012), wine (e.g. Chiriacò et al. 2019; Marchi et al. 2018) and coffee (e.g. Birkenberg and Birner 2018; Walenta 2015).

Nonetheless, only two of these Kilian et al. (2012) and Birkenberg and Birner (2018) are applicative case studies of carbon neutral principles. The two case studies that we will present in the following sections are based in Costa Rica. This small nation has indeed committed itself to become the first carbon neutral country by 2021 by developing a governmental strategy towards carbon neutrality with an action orientation (Carrasquilla-Batista et al. 2017).

6.3.1 *Carbon Neutral Fruits: The Dole Carbon Neutral Strategy for Bananas and Pineapples*

The case study of Dole company carbon neutral strategy has been explored in literature by Kilian et al. (2012). The authors investigate the actions implemented by the largest fruit and vegetable producer on the planet to reduce and mitigate the emissions associated with their products. In 2007 Dole announces its commitment to become a carbon neutral company by firstly offering to the market carbon neutral bananas and pineapple (Fig. 6.11).

The first step of the company was to elaborate with the advice of Soil & More International B.V. and certified by TUEV-No (according with the PAS 2050 and the ISO standard 14,044) the carbon footprint of their bananas produced on Dole-owned banana plantations in Costa Rica and then distributed in the German marketplace. The results indicate that the carbon footprint of the bananas amounts to 1124 kg CO2e per ton of banana (Soil & More International B.V. 2010)(Fig. 6.12).

Once they had calculated their carbon footprint and found the negative hotspots of their production, they implemented a series of measures to mitigate the emissions associated with the whole life cycle of their bananas and their company operations (Dole 2019). The following actions have been retrieved by the company website:

- Control Released Fertilizer Program and Precision Agriculture: assessed specific fertilization needs and introduced new Control Released Fertilizer product (fertilizer use decreased by 12% and GHG emissions by 45%).
- Driving Techniques Training: They trained employees on better driving techniques (fuel consumption -10%).
- Railroad Transportation: 40% of the bananas produced are transported by train (34% decrease in fuel consumption and GHG emissions)
- Fuel Conservation: fuel quotas and fuel consumption tracking systems on its heavy equipment vehicles as a way to continuously measure performance.

Fig. 6.11 Dole logo
Source: www.dolecrs.com

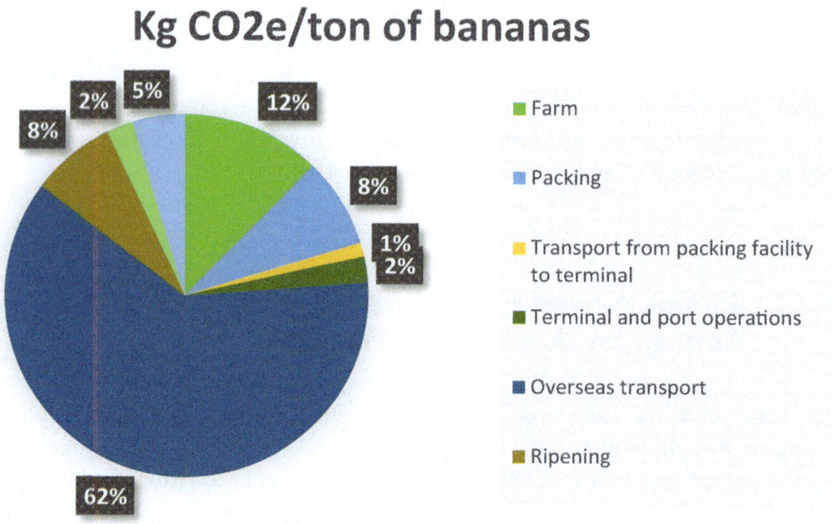

Fig. 6.12 Dole's bananas Carbon footprint
Source: Soil & More International B.V. (2010)

- Biodigesters: At Bananito farm, biodigesters have been used to t convert the mules' feces into gas that powers the workers' cafeteria

Moreover, in container yard, Dole reduced between 2007 and 2009 its use of fuel by 33% through for example the introduction of control mechanisms for container handling equipment, establishing fuel quotas and the implementation of reduction goals. In the container yard and container transportation, Dole reduced its greenhouse gas emissions by over 90% in 5 years due to: preventative maintenance and repair programs, replacement of refrigeration units for more modern equipment, new units with superior insulation and compressor systems, measurement and certification of the CO2e emissions of our vessels. Additionally, the company implemented carbon reduction actions also in their headquarters in California, as for example with the use of energy efficient lamps, lighting sensor, 5-Star energy rating system for meeting air filtration standards and incentives for employees who participate in the commuter program.

Furthermore, Dole is working whit FONAFIFO (Fondo Nacional de Financiamento Forestal), a Costa Rican Government forestry agency, to offset the CO_2 emissions related to the transportation of their bananas and pineapples by planting trees and protecting existing forest. In Kilian et al. (2012) three main motivations for Dole's carbon neutral strategy are cited: a carbon neutral strategy has been seen by the company as a natural extension of their environmental programs, customer inquiries about the company's carbon footprint and plans and the Costa Rican Government's commitment to carbon neutrality by 2021.

6.3.2 *Carbon Neutral Coffee: The Coopedota Case Study*

This second case study has been reported in literature by Birkenberg and Birner (2018), Jiménez et al. (2013) and Castro (2015). The case study is about the Coopedota a Costa Rican coffee cooperative of 800 associated coffee farmers, that was the first one to be certified carbon neutral based on a widely recognized international standard, the PAS 2060 (Fig. 6.13).

Coopedota was yet a pioneer in sustainable agricultural practices and their coffee was already certified according to Rainforest Alliance, Coffee Practice of Starbucks, ISO 9001 and 14,001 and Fair Trade. The idea of carbon neutral certification was launched in 2008 also as a consequence of the Costa Rica government commitment to carbon neutrality and by 2011 they certified their first coffee. Jiménez et al. (2013) investigate the steps that lead Coopedota to the carbon neutral certification (Fig. 6.14).

In particular, Birkenberg and Birner (2018) reported the amount of coffee certified over the years, the carbon footprint of coffee and the total of the emission to offset (Table 6.2).

They also perform a very interesting Social Network Analysis and 'Process Net-Map' to visualize the social networks that enabled Coopedota to pursue the certification scheme (Birkenberg and Birner 2018).

Additionally, Birkenberg and Birner (2018) highlighted some challenges encountered by Coopedota in achieving the certification:

- **Farm management data:** retrieving this type of data was one of the main challenges for Coopedota, mainly due to small farmers inability to collect this data. The company overcome this difficulty engaging farmers already involved with Rainforest Alliance project, so more used to collect farm data.
- **Carbon offsetting mistrust:** generally offsetting practices are seen by consumers as "greenwashing".

Fig. 6.13 Coopedota carbon neutral initiative
Source: www.coopedota.com

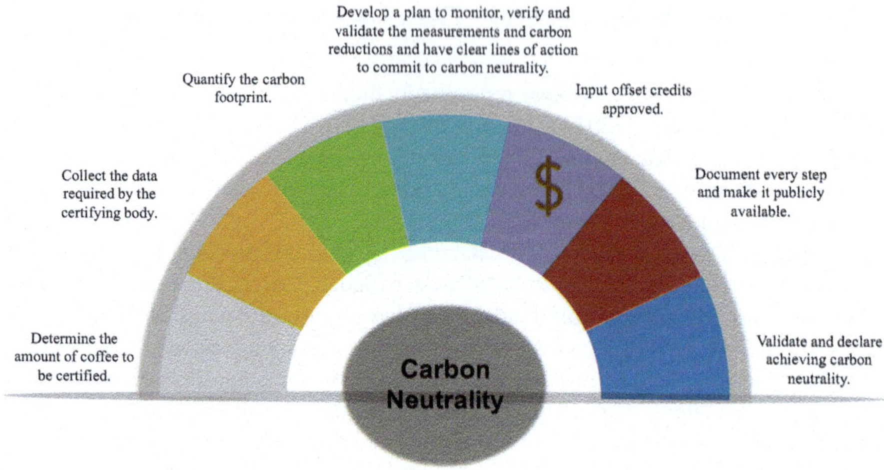

Fig. 6.14 Coopedota steps towards carbon neutrality
Source: Authors own elaboration based on Jiménez et al. (2013)

Table 6.2 Coopedota amount of certified coffee, the carbon footprint and emission to be offset

Years	Type of coffee	2010–2011	2011–2012	2012–2013	2013–2014	2014–2015
Amount of coffee certified (kg green coffee)	Green coffee	920,000	460,000	460,000	460,000	460,000
	Roasted coffee	23,000	23,000	23,000	23,000	23,000
Emissions (t CO^2eq)	Green coffee	1800	917	1443	508	859
	Roasted coffee	69	62,	90	48	82
Carbon Footprint (kg CO^2eq $kg^{-}1$ green coffee) Total	Green coffee (cradle-to-gate)	1,96	1,99	3,14	1,10	1,87
	Roasted coffee (cradle-to-grave)	3,00	2,70	3,92	2,10	3,15
Total emissions that need to be offset b (t CO^2eq)	Green & roasted coffee	1869	979	1533	556	940

Source: Birkenberg and Birner (2018)

- **Cost:** the certification present cost associated with both emissions' reduction, data collection, certification fees and offsetting (see Table 6.3). However, the certification also present opportunities for cost saving as: reduced costs for energy and fertilizers, price premium and positive image.

Table 6.3 Internal and external certification cost

Internal cost budget of Coopedota for carbon neutral certification (US$)			Budget for external costs for carbon neutral certification (US$)		
Internal cost	**Certification 2011**	**Review 2012**	**External Cost**	**Certification 2011**	**Review 2012**
Travel and per diem	2105.0	–	**Certification (audit)**	15,000	4050
Labor	8700.0	2000.0	**Carbon credits**	9400	8442
Other expenses	150.0	–	**Professional services**	4600	
Coopedota's workforce	8203.2	1871.6	–		
Total	19158.2	3871.6	**Total**	29,000	12,492

Source: Department of finance, Coopedota RL in Jiménez et al. (2013)

- **Difficulties related to CF calculation:** CF presents some difficulties in comparison as functional and system boundaries (cradle-to-gate or cradle-to-grave) unit can differ.
- **Marketability:** As we stated also above in this report, carbon neutrality presents some challenges in being communicated to consumers as it often overlaps with other certifications, it does not present information on the label regarding GHG emissions generated, reduced and offsetted. Moreover, offsetting is not always seen positively by consumers.
- **Carbon sequestration:** PAS 2060 does not accounts for on-site carbon sequestration. Nonetheless, on-site carbon sequestration is a great opportunity for perennial corps as coffee and can lead to reduced cost for offset.

6.3.3 Carbon Neutral Meat: The Five Founders Case Study

Five Founders beef is produced by the North Australian Pastoral Company (NAPCO), one of Australia's largest privately-owned cattle companies. NAPCO currently manages over 6.1 million hectares, in both Queensland and the Northern Territory. NAPCO also operates the Wainui Feedlot and Farm on the Darling Downs. Combined, this property portfolio supports approximately 200,000 head of cattle. In April 2019 Five Founders became Australia's First Carbon Neutral certified beef. They are certified by the Australian Government with the Climate Active program (ex NCOS) (Fig. 6.15).

According to the Australian Climate Active standard their carbon neutrality strategy consists in three phases: Assess, reduce, offset. The main characteristic of these phases for the Five Founders meat brand are synthetizes in Fig. 6.16.

Fig. 6.15 Five Founders Carbon Neutral logo
Source: https://fivefounders.com.au/our-difference/#environment

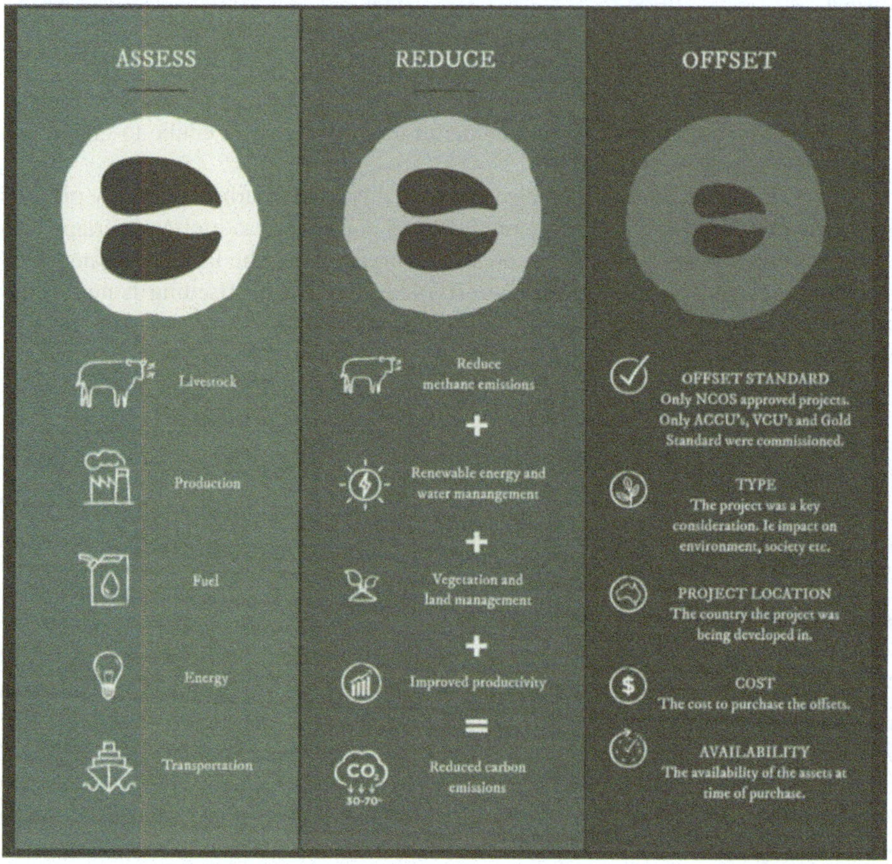

Fig. 6.16 Five Founders Carbon Neutrality strategy
Source: https://fivefounders.com.au/our-difference/#environment

The Assessment of the Carbon Fooprint

For assess their Carbon Footprint Five Founders choose as functional unit 20 kg branded beef package. A hybrid LCA methodology was used. Livestock emissions were calculated using a herd inventory and emission calculation approach derived from the ERF Beef Herd Management method, combined with estimation of energy related emissions from the farm and feedlot. This was combined with data from input-output analysis (based on $ value of expenditure) to calculate other (off-farm) supply chain inputs. Input-output analysis provides a complete assessment of the downstream inputs that goes into the delivery of branded beef products to the wholesaler/distributor warehouse. They adopted a Cradle to Gate boundary approach that includes all activities involving cattle raising, feeding, transporting, slaughtering, product packaging, distribution to customer/wholesaler and chilled storage. Activities involved is retailing of products, storage in home refrigeration, cooking and consumption of beef by consumers has not been included (Fig. 6.17).

Results from the LCA analysis reported the following emissions for the base year 2018 calendar year (January–December) reporting period (Fig. 6.18).

NAPCO calculated its carbon footprint to understand its current emissions profile and to inform the development of an emissions reduction strategy (ERS). Emissions from enteric methane clearly dominate the footprint and it is for this reason the company will focus on initiatives that reduce emissions associated with the livestock. In particular, the greatest opportunity comes with the breeder herd which is responsible for largest portion of the enteric emissions. This is where NAPCO will focus its initial efforts for the ERS to deliver the largest emissions reduction. NAPCO is in the early stage of its emissions reduction strategy with initiatives to date including: replacing diesel powered bores with solar power, feed efficiency to increase live weight gain, participating in legume trials to measure potential methane reduction and increasing the density of watering points to increase grazing radius which contributes to reduced time to slaughter. Moving forward the impact of these initiatives along with additional ones that have been identified will be formally measured to understand the emission reduction benefit.

NAPCO Offset Strategy NAPCO is supporting Australian regenerative vegetation offset projects issued by the Australian Government as well as renewable energy projects in China certified by VCS (Fig. 6.19). Offsets are forward purchased based on the previous assessment for the current quarters estimated shipping volume initially starting with April 1st 2019. At the start of the next quarter a true-up will occur and any additional credits will be procured. At the same time offsets are purchased for the current quarter's estimated shipments. If any quarter is in surplus, then it is carried over to the next quarter.

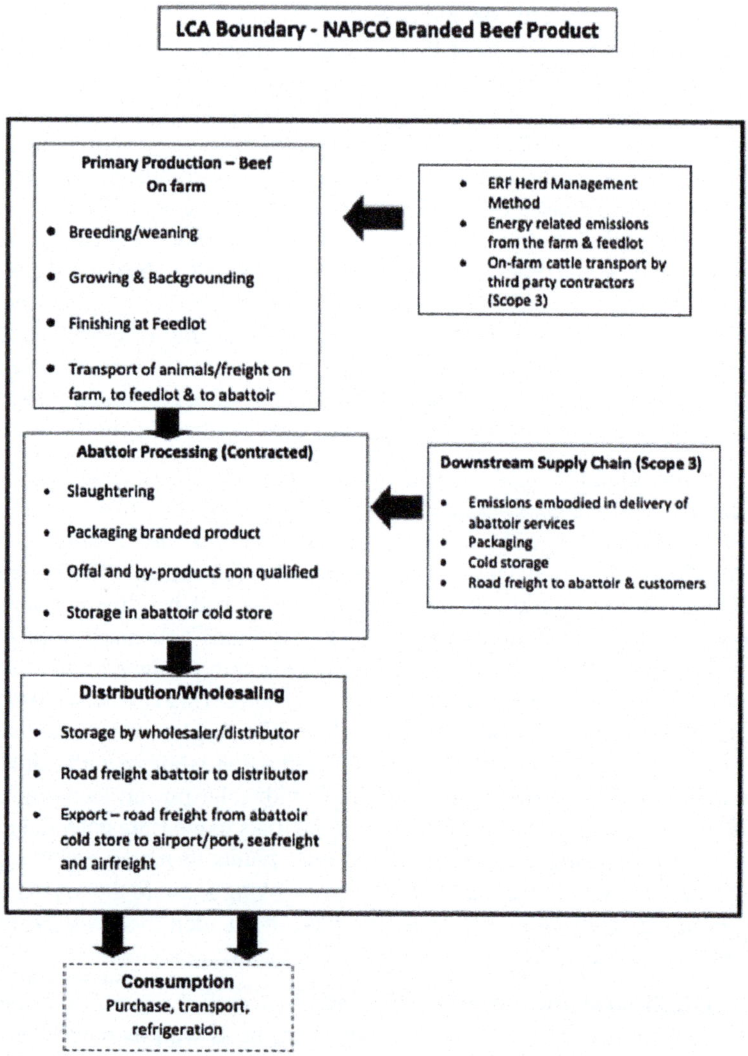

Fig. 6.17 Five Founders LCA Boundary
Source: NAPCO' Australian Government Carbon Neutral Program Public Disclosure Summary (2019)

Table 1. Base Year Emissions January – December 2018 (tCO_2-e)	
Scope 1	7,869.89
Scope 2	4.93
Scope 3	1,278.75
Total	9,153.57

2. Emissions summary

Scope	Emission source	tCO2-e
2	Feedlot electricity	4.93
1	Feedlot stationary fuel (diesel)	9.51
1	Feedlot LPG	8.41
3	Feedlot rations	245.03
1	On farm stationary diesel	120.50
1	On farm petrol (post 2004)	3.86
1	On farm aviation fuel	7.12
1	On farm LPG	1.15
3	On farm LPG	0.36
1	Livestock emissions	7,719.34
3	On farm road freight	258.48
3	On farm feeding supplements	31.44
3	Feedlot electricity	0.80
3	Feedlot stationary fuel diesel	0.49
3	Feedlot LPG	0.50
3	On farm stationary diesel	6.18
3	On farm petrol (post 2004)	0.21
3	On farm aviation fuel	0.38
3	Road freight farm to abattoir	96.14
3	Abattoir scope 1 emissions slaughtering	169.55
3	Abattoir Scope 2 emissions from slaughtering	139.05
3	Abattoir slaughter services	117.91
3	Abattoir packaging materials	25.29
3	Cold Storage abattoir	5.70
3	Domestic chilled freight Abattoir to distributor warehouse	145.25
3	Domestic chilled storage	1.00
3	Export - Airfreight	16.77
3	Export - Sea freight	18.22
	Total (based on initial forecast of product expected to be sold in 2019)	9,153.47

Fig. 6.18 Emissions of Five Founders (2018)
Source: NAPCO' Australian Government Carbon Neutral Program Public Disclosure Summary (2019)

Table 3. Offsets Summary			
Offset type and registry	Date cancelled	Quantity	Serial numbers
Australian Carbon Credit Unit (ACCU), Australian Voluntary Cancellation Account Vintage: 2018-19 Project: Paroo River South Environmental Project – Human Induced Regeneration (HIR)	29 March 2019	62	3,779,588,756 - 3,779,588,817
Verified Carbon Units (VCUs) - CECIC HKC Danjinghe Wind Farm Project, China, APX VCS Registry Vintage: 29/12/2015 to 31/12/2015	29 March 2019	2063	6498-324162616-324164678-VCU-034-APX-CN-1-1839-29122015-31122015-0
Verified Carbon Units (VCUs) Rice Husk Based Thermal Energy Generation Project at Thot Not, China, APX VCS Registry Vintage: 01/03/2013 to 28/02/2014	29 March 2019	375	3983-170784173-170784547-VCU-008-APX-VN-1-908-01032013-28022014-0
Total offset units cancelled			2,500
Net emissions after offsetting			0

Fig. 6.19 NAPCO offset strategy
Source: NAPCO' Australian Government Carbon Neutral Program Public Disclosure Summary (2019)

References

Audsley E, Brander M, Chatterton J, Murphy-bokern D, Webster C, Williams A (2009) How low can we go? An assessment of greenhouse gas emissions from the UK food system and the scope to reduce them by 2050. WWF-UK and Food Climate Research Network. https://doi.org/10.1093/bja/aeh529

Austgulen MH, Skuland SE, Schjøll A, Alfnes F (2018) Consumer readiness to reduce meat consumption for the purpose of environmental sustainability: insights from Norway. Sustain. https://doi.org/10.3390/su10093058

Barilla Center for Food and Nutrition (2016) DoUBLE PYRAMID 2016 Eat better Eat less Food for all

Barners-Lee M (2012) Booths greenhouse gas footprint report

Birkenberg A, Birner R (2018) The world's first carbon neutral coffee: lessons on certification and innovation from a pioneer case in Costa Rica. J Clean Prod 189:485–501. https://doi.org/10.1016/j.jclepro.2018.03.226

BSR (2009) Value chain approaches to a low-carbon economy: business and policy partnerships a discussion paper for the world business summit on Climate Change in Copenhagen

Carrasquilla-Batista A, Chacon-Rodriguez A, Solorzano-Quintana M, Guerrero-Barrantes M (2017) IoT applications: on the path of Costa Rica's commitment to becoming carbon-neutral. Internet Things Glob. Community, IoTGC 2017 – Proc. https://doi.org/10.1109/IoTGC.2017.8008975

Castro R (2015) Eco-competitiveness and eco-efficiency: carbon neutrality in Latin America. Harvard Proj Clim Agreements:1–12. https://doi.org/10.1787/888933272500

Cederberg C, Meyer D, Flysjö A (2009) Life cycle inventory of greenhouse gas emissions and use of land and energy in Brazilian beef production, SIK Report No 792

Chiriacò MV, Belli C, Chiti T, Trotta C, Sabbatini S (2019) The potential carbon neutrality of sustainable viticulture showed through a comprehensive assessment of the greenhouse gas (GHG) budget of wine production. J Clean Prod 225:435–450. https://doi.org/10.1016/j.jclepro.2019.03.192

Climate-Smart Agriculture Leadership Forum (2016) A climate-smart pathway for Irish agricultural development exploring the leadership opportunity

Deloitte (2016) Deloitte Sustainability. Circular economy potential for climate change mitigation, pp 0–43

Dole (2019) Carbon footprint reduction [WWW document]. URL http://dolecrs.com/sustainability/carbon-footprint/reduction/. Accessed 8.21.19

Elofsson K, Bengtsson N, Matsdotter E, Arntyr J (2016) The impact of climate information on milk demand: evidence from a field experiment. Food Policy. https://doi.org/10.1016/j.foodpol.2015.11.002

European Commission (2006) Environmental impact of products (EIPRO): analysis of the life cycle environmental impacts related to the Total final consumption of the EU 25 (2006). European Commission technical report EUR 22284 EN

FAO (2011) Energy-smart food for people climate

FAO (2018) FAO'S work on climate change

Fischedick M, Roy J, Abdel-Aziz A, Acquaye A, Allwood JM, Ceron J-P, Geng Y, Kheshgi H, Lanza A, Perczyk D, Price L, Santalla E, Sheinbaum C, Tanaka K (2014) Industry, in: Limate change 2014: mitigation of climate change. Contribution of working group III to the fifth assessment report of the Intergovernmental Panel on Climate Change. Cambridge University Press, Cambridge/New York

FoodDrinkEurope (2015) A TIME TO ACT Climate Action and the Food and Drink Industry

Garnett T (2008) Cooking up a storm and our changing climate, strategy

Garnett T (2011) Where are the best opportunities for reducing greenhouse gas emissions in the food system (including the food chain)? Food Policy. https://doi.org/10.1016/j.foodpol.2010.10.010

Gladek, E., Kennedy, E., Exter, P. van, Monaghan, C., 2018. Food value chains strategy review

Hartikainen H, Roininen T, Katajajuuri JM, Pulkkinen H (2014) Finnish consumer perceptions of carbon footprints and carbon labelling of food products. J Clean Prod 73:285–293. https://doi.org/10.1016/j.jclepro.2013.09.018

Jiménez GA, Kilian B, Rivera L (2013) Sustainability in the coffee growing business : Coopedota and the path towards carbon neutral coffee

Kilian B, Hettinga J, Jiménez GA, Molina S, White A (2012) Case study on Dole's carbon-neutral fruits. J Bus Res 65:1800–1810. https://doi.org/10.1016/j.jbusres.2011.10.040

KPMG, 2013. The agricultural and food value chain: entering a new era of cooperation

Leire C, Thidell Å (2005) Product-related environmental information to guide consumer purchases – a review and analysis of research on perceptions, understanding and use among Nordic consumers. J Clean Prod. https://doi.org/10.1016/j.jclepro.2004.12.004

Marchi M, Neri E, Pulselli FM, Bastianoni S (2018) CO2 recovery from wine production: possible implications on the carbon balance at territorial level. J CO2 Util 28:137–144. https://doi.org/10.1016/j.jcou.2018.09.021

Mayberry D, Bartlett H, Moss J, Davison T, Herrero M (2019) Pathways to carbon-neutrality for the Australian red meat sector. Agric Syst 175:13–21. https://doi.org/10.1016/j.agsy.2019.05.009

Meier T, Christen O (2012) Gender as a factor in an environmental assessment of the consumption of animal and plant-based foods in Germany. Int J Life Cycle Assess. https://doi.org/10.1007/s11367-012-0387-x

Muller A, Jawtusch J, Gattinger A (2011) Mitigating greenhouse gases in agriculture. Brot für die Welt 88

Murphy-bokern D, Kleemann L (2014) The role of corporate social responsibility in reducing greenhouse gas emissions from agriculture and food

Nieberg H (2009) Auf den Nahrungskonsum zurückzuführende THG-Emissionen., in: Erfassung, Bewertung Und Minderung von Treibhausgasemissionen Des Deutschen Agrar- Und Ernährungs- Sektors. Arbeitsberichte Aus Der VTI-Agrarökonomie 03/2009. Braunschweig: VTI

Niles, M., Esquivel, J., Ahuja, R., Mango, N., 2017. Climate Change & Food Systems : assessing impacts and opportunities

Niles MT, Ahuja R, Barker T, Esquivel J, Gutterman S, Heller MC, Mango N, Portner D, Raimond R, Tirado C, Vermeulen S (2018) Climate change mitigation beyond agriculture: a review of food system opportunities and implications. Renew Agric Food Syst. https://doi.org/10.1017/S1742170518000029

Noleppa S (2012) Climate change on your plate. World Wild Fund Nat. Ger

Peters GM, Rowley HV, Wiedemann S, Tucker R, Short MD, Schulz M (2010) Red meat production in Australia: Life cycle assessment and comparison with overseas studies. Environ Sci Technol. https://doi.org/10.1021/es901131e

Poore J, Nemecek T (2018) Reducing food's environmental impacts through producers and consumers. Science (80-.). 360, 987–992. https://doi.org/10.1126/science.aaq0216

Retail Forum for Sustainability (2011) Measurement and reduction of carbon footprint of stores, pp 1–10

Sharp A, Wheeler M (2013) Reducing householders' grocery carbon emissions: carbon literacy and carbon label preferences. Australas Mark J 21:240–249. https://doi.org/10.1016/j.ausmj.2013.08.004

Sim S, Barry M, Clift R, Cowell SJ (2007) The relative importance of transport in determining an appropriate sustainability strategy for food sourcing. A case study of fresh produce supply chains. Int J Life Cycle Assess. https://doi.org/10.1065/lca2006.07.259

Smith P, Bustamante M, Ahammad H, Clark H, Dong H, Elsiddig EA, Haberl H, Harper R, House J, Jafari M, Masera O, Mbow C, Ravindranath NH, Rice CW, Abad CR, Romanovskaya A, Sperling F, Tubiello F (2014) Agriculture, forestry and other land use (AFOLU). In: Climate Change 2014: mitigation of climate change. Contribution of Working Group III to the Fifth Assessment Report of the Intergovernmental Panel on Climate Change. Cambridge University Press, Cambridge/New York, pp 811–922

Soil & More International B.V (2010) Comprehensive Carbon Footprint Assessment Dole Bananas

Stoessel F, Juraske R, Pfister S, Hellweg S (2012) Life cycle inventory and carbon and water foodprint of fruits and vegetables: application to a swiss retailer. Environ Sci Technol. https://doi.org/10.1021/es2030577

Styles D, Schoenberger H, Galvez-Martos JL (2012) Environmental improvement of product supply chains: a review of European retailers' performance. Resour Conserv Recycl 65:57–78. https://doi.org/10.1016/j.resconrec.2012.05.002

Vanclay JK, Shortiss J, Aulsebrook S, Gillespie AM, Howell BC, Johanni R, Maher MJ, Mitchell KM, Stewart MD, Yates J (2011) Customer response to carbon labelling of groceries. J Consum Policy 34:153–160. https://doi.org/10.1007/s10603-010-9140-7

Walenta J (2015) Becoming carbon neutral: evaluating the carbon neutral certification as a tool for reducing climate change impacts and securing financial livelihoods. Sustain J Rec 8:121–126. https://doi.org/10.1089/sus.2015.29002

Weber CL, Matthews HS (2008) Food-miles and the relative climate impacts of food choices in the United States. Environ Sci Technol. https://doi.org/10.1021/es702969f

Zimmermann A, Benda J, Webber H, Jafari Y (2018) Trade, food security and climate change: conceptual linkages and policy implications

Chapter 7
Carbon Neutrality as Business: Decoupling Growth from Emissions While Managing Climate Risks

The chapter discuss carbon neutrality as a marketing tool for brand building, pricing, willingness to pay, and developing new market segments. Carbon neutrality is compared with other greening themes that can be marketed to consumers. Additionally, the chapter focus on final consumers but also on B2B and other possible targets such as access to green public procurement. Then, the chapter discusses in detail different market drivers for companies to move towards lower carbon emissions and carbon neutral process and products. Furthermore, Barriers and risks for suppliers to join the carbon neutral chain have been analysed. Finally, it summarizes the main pathways that companies are taking in terms of marketing, relative to carbon neutrality and what this means in terms of trends in the market.

7.1 The Role of consumers' Food Choices in Mitigating Carbon Emissions

Consumers can play a pivotal role in reducing GHG emissions, shifting their consumption pattern towards more sustainable and low carbon choices. Mason and Lang (2017) argued that only changes in the food production system would not be enough to ensure long term sustainability of the sector, consumers dietary patterns need to change too. In the plenary session of the FAO Symposium "Biodiversity and Sustainable Diets united against hunger" a definition of Sustainable Diets has been proposed as: *"those diets with low environmental impacts which contribute to food and nutrition security and to healthy life for present and future generations. Sustainable diets are protective and respectful of biodiversity and ecosystems, culturally acceptable, accessible, economically fair and affordable; nutritionally adequate, safe and healthy; while optimizing natural and human resources"*(FAO 2010).

A. Acampora et al., *Carbon Neutrality in the Agri-food Sector*,
https://doi.org/10.1007/978-3-030-88048-4_7

In this sense Garnett (2014) founded synergies between healthier and climate-friendly diets and has listed their main characteristics:

- Diversity: a wide variety of foods eaten
- Balance achieved between energy intake and energy needs
- Based around: minimally processed tubers and whole grains; legumes; fruits and vegetables – particularly those that are field grown, 'robust' (less prone to spoilage) and less requiring of rapid and more energy-intensive transport modes
- Meat eaten sparingly if at all – and all animal parts consumed
- Dairy products or alternatives eaten in moderation e.g. fortified milk substitutes and other foods rich in calcium and micronutrients
- Unsalted seeds and nuts
- Small quantities of fish and aquatic products sourced from certified fisheries and certified aquaculture systems
- Very limited consumption of processed foods high in fat, sugar or salt and low in micronutrients e.g. crisps, confectionery, sugary drinks

However, changing consumers consumptions patterns toward more sustainable diets would imply culture and social consideration. Food serve not only to provide nourishment but also to provide pleasure heavily influenced by social and cultural traditions and religious beliefs (Johnston et al. 2014). Several studies support the assumption that changing consumers' diet (towards vegan or vegetarian) (Faber et al. 2012; Garnett 2011) and their behavior toward low carbon or carbon neutral food can significantly reduce GHG emissions at food production and consumption levels (Lombardi et al. 2017). At the same time, consumers concern about environmental issues and climate change has increased sharply. Moreover, the Greta Thunberg international youth movement against climate change "#FridaysforFuture" has stirred the consciences of citizens and has encouraged global awareness of the negative consequences of climate change. Several scholars have found that consumers are increasingly concerned about their impacts on the environment and on the climate (Sirieix et al. 2013; Thøgersen and Nielsen 2016; Vanclay et al. 2011). However, these environmental concerns are difficult to translate into actual consumers' behavioral change (Iweala et al. 2019). This is particularly true for food choices. Even if groceries and food choice represent for consumers a great opportunity to exert high degree of personal choice making more greener 'day-to-day' choices and to reduce their carbon footprint (Hartikainen et al. 2014; Hertwich 2005), the speed, the automatism and repetitiveness of their products choices at the shelf represent a barrier towards behavior change (Sharp and Wheeler 2013). However, altering consumers consumption and buying patterns for food products would imply providing consumers with accurate information about environmental problems and the environmental impacts of their purchasing choices (Leire and Thidell 2005). Camilleri et al. (2019) found that American consumers are often unaware or misinformed about energy consumption and GHG emission of different food products, systematically underestimating the environmental impacts of different food items. According to this study, the familiarity with these products that consumers found at the shelf and, at the same time, the "invisibility" and complexity of the production and distribution of food result in a generally misinformation about the emissions associated with these products (Camilleri et al. 2019).

7.2 Consumers' Willingness to Pay for Climate Friendly and Climate Neutral Products

Given the cost associated with implementing labelling schemes it is crucial for companies to explore if they can charge low carbon products with a premium price. Several studies investigated consumers' willingness to pay (WTP) for low or neutral carbon food products (Table 7.1). However, findings from these studies are not straightforward, leading to different results (Onozaka et al. 2016).

Several studies' evidence suggest that most consumers are willing to pay more for a food product with a lower carbon impact (Caputo et al. 2013; De Marchi et al. 2016; Onozaka and McFadden 2011; Shuai et al. 2014). Notably, in their study Ma et al. (2018) found that while 91% of respondents are willing to buy low carbon vegetables in Shanghai, their willingness to pay extra for them is limited. Consumers are, indeed, willing to pay +6% in comparison to the traditional product. Van Loo et al. (2014), found that Belgian consumers are willing to pay a price premium of up to 24% for carbon-labeled chicken breast. Chen et al., (2018) explored Chinese consumers' willingness to pay for purified bottled water with carbon labels. They found that consumers are willing to pay consistently more for carbon labelled water and that WTP decreased significantly when the carbon emissions of purified bottled water increased. Additionally, Kima et al. (2016), suggested that the impact of

Table 7.1 Synthesis of the results from studies investigating WTP for carbon labeled products

Authors	Type of label	Food product	Country	WTP
Ma et al. (2018)	Low carbon	Vegetables	Shanghai (China)	+6%
Van Loo et al. (2014)	Carbon labelled	Chicken breast	Belgium	+24%
Chen et al. (2018)	Carbon label	Purified bottled water	China	More
Drichoutis et al. (2016)	Carbon neutral	Eggs and oil	Greece	+27.6% (eggs); +23.3% (oil)
Lombardi et al. (2017)	Climate neutral	Milk	Italy	+ 0.55 €/liter
Gassler (2014)	Carbon neutral	Wine	Austria	+ € 1.43

Source: Authors' own elaboration

climate change on consumers personal lives is an important indicator to predict the success of carbon labels in the market. Findings from their study highlighted that South Korean consumers more exposed to the risk of environmental disorder or living in a vulnerable area to climate change were more willing to pay a premium price for low carbon emission labelled apples (Kima et al. 2016). Other studies, instead, found that consumers are unwilling to pay more for carbon labelled foods (Edwards-Jones et al. 2009; Grunert et al. 2014; Sorensen 2009; Upham and Bleda 2009) and that carbon labels do not have a strong impact on their buying decisions (Hartikainen et al. 2014). However, the majority of the studies found that consumers' willingness to pay for climate friendly food products is generally low especially in comparison with other green labels (e.g. organic, locally produced etc.) (Feucht and Zander 2018; Gadema and Oglethorpe 2011; Hartikainen et al. 2014; Iweala et al. 2019; Van Loo et al. 2014). Furthermore, Drichoutis et al. (2016) explored consumer acceptance and estimated the premium associated with climate neutral foods founding that Greek consumers are willing to pay, on average, premiums of up to 27.6% and 23.3% for climate neutral eggs and olive oil, respectively. Also Lombardi et al. (2017) investigated the premium price Italian consumers are willing to pay for climate neutral milk. They found that consumers are willing to pay 0.55 €/liter more for climate neutral milk and that this predisposition to pay a premium price is higher after being exposed to the carbon label information (0.57 €/liter). Contrary, Stokes and Turri (2015), in their study on American consumers' perception of different carbon neutral labelled products (ketchup and household insecticide spray), demonstrated that even if the presence of a carbon-neutral label leads to more favorable perceptions of company environmental concern, the labels do not influence company credibility nor purchase intentions. Additionally, Pomarici and Vecchio (2014) findings shows that the Italian "millennials generation" is very interested in buying carbon neutral labeled wine (almost 65% of their sample stated that they are willingness to buy climate neutral wine) even if the knowledge about this label is very low. Furthermore, Gassler (2014) found that Austrian consumers are willing to pay € 1.44 more for sustainable wine and € 1.43 more for carbon neutral labeled wine. Also Iweala et al. (2019) in their study on UK and German consumers' attitudes on prosocial (fair-trade) and pro-environmental (organic and carbon-neutral) claims for chocolate, highlighted that awareness about the climate-neutral label among consumers is low (in their sample is below 20%). They assumed that the "warm glow of giving", the act of "feeling people experience when performing an apparent altruistic act" does not influence the choice of carbon neutral products, like when they chose organic or fair trade. They stated that this is not because people do not care about the damage provoked by climate change but because the lack of awareness about the climate neutral label does not make them feel like they are doing something about climate change (Iweala et al. 2019).

7.3 Market drivers and Businesses Opportunity for Agri-Food Companies to Move Towards Lower Carbon Emissions

The pathway towards low carbon transition cannot be completed without the support of companies (Kolk and Pinkse 2005). These, in fact, play a fundamental role in the strategy towards the reduction of GHG emissions without which the objectives of the Paris agreement will be more difficult to achieve (World Economic Forum 2018). Business engagement in climate friendly activities is, therefore, the focus of emission reduction policies both nationally and internationally. Fortunately, companies are progressively taking the lead in guiding this low carbon transition because they have also start to see this strategy as a way to save costs and increase profit (Bartlett et al. 2016). However, business take action for different reasons, but they need to be certain that a low carbon pathway makes good business sense (We Mean Business 2015). The aim of this section is to summarize the main market drivers explored in scholars' literature and in companies, government and NGO reports.

For instance Pinkse and Busch (2013) elaborated the framework showed in Fig. 7.1 in which they summarized the main factors that impact a firm's choice to adopt a specific corporate carbon norm. They analyzed the strategic factors (image

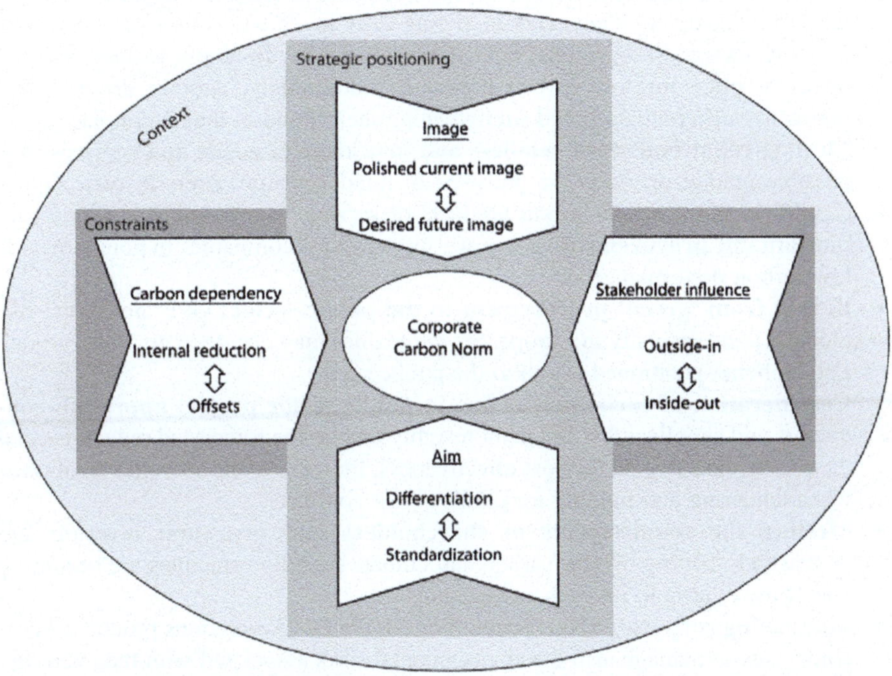

Fig. 7.1 Pinkse and Busch (2013) framework
Source: Pinkse and Busch (2013)

and aim) and constraint (carbon dependency and stakeholder influence) that firms encountered in their carbon norms implementation. Strategic factors are image, they can chose to polish their current image or to project on the market a new green desired future image, and aim, they can chose to apply carbon norms to differentiate themselves from competitors or to reach the level of carbon norms of the market and thus standardize their production at the level of their competitors (Pinkse and Busch 2013).

Moreover, in a report by InterTraderIreland (2011) they suggest the following drivers and business benefits for companies to manage and report their GHG emissions:

- **Emissions can act as an indicator of resource efficiency:** as GHG emissions are strictly correlated with resource consumption (especially energy consumption) they can serve as useful indicator of overall business efficiency. In this sense carbon management can serve to track 'carbon hotspots' or efficiency gaps, and thus identify corrective actions to apply.
- **Demonstrating compliance with carbon-related regulations**: all over the world governments are implementing national targets for GHG emissions reduction, these targets have also implication for company. In this sense developing a carbon protocol can help firms to stay in line with the regulations and comply with national and international policy.
- **Reputational improvement and differentiation in the marketplace:** companies can decide to apply carbon norms also in consideration of consumers growing awareness on climate change issues and willingness to buy and pay premium price for low carbon products. Additionally, another driver is the possibility differentiating and strengthening their brand in the marketplace.
- **Meeting requirements of retailers and procurers of goods and services:** one possible market driver is the presence of retailers or procurers in their supply chain that requires low carbon products or service. The possibility to profit by "**sustainable procurement**" is a powerful driver for companies in going towards low carbon performance.
- **Benefit from 'green' procurement in the public sector:** GPP are on the rise globally and especially in Europe. So, companies may decide to go "low carbon" also be better positioned to sell to the public sector.
- **Consideration of carbon reduction in public sector project appraisal:** contractors and suppliers that can demonstrably deliver the forecasted reductions and the pre-defined targets for cost effectiveness, may therefore be better positioned when tendering and bidding for public sector contracts.
- **Meeting the requirements of shareholders and investors:** investors are increasingly betting on low carbon and climate friendly principles are becoming one of the criteria in investment decisions.
- **Minimizing corporate risks:** accounting for the GHG emissions produced helps companies to manage also the environmental risks associated with their activity. Minimizing risks regards also the ability to reduce the risk of possible future regulation that taxes carbon consumption.

Furthermore, Bartlett et al. (2016) added other two drivers to bear in mind:

- **The possibility of getting ahead of policy**: being a precursor in the field may lead pioneering companies to the possibility of shaping the regulation on the topic.
- **Securing market position:** Businesses that are taking action will also be among the best situated to take advantage of the considerable new opportunities in low-carbon goods and services.

Murphy-bokern and Kleemann (2014) pointed out another driver for companies:

- **Partnership:** Corporate initiatives, roundtables and workshop have been organized around the world to fight climate change and to develop guidelines and suggestions to lower companies' emissions. These initiatives have created opportunities for partnerships between firms to achieve the shared goal of reduce GHG emissions. In particular, in the food sector different commodity roundtables have been created with the aim of manage certification schemes, spreading agricultural good practices and certified product or services among participants.

Some authors argue that one of the main drivers to go net-zero emissions for agri-food companies is to protect their business (Changing Markets Foundation 2018; Corbett 2009; Lóránt and Allen 2019).

- **Future-proof the organization**: Agriculture is among the most exposed sectors to climate change. Farming is strictly dependent on the natural environment and changes in rainfall and increasing temperature extremes are reported as significant risks across all the supply chain. To implement a carbon strategy will defend companies to the adverse consequences of climate change on agriculture.

Moreover, the association We Mean Business (2015) a coalition of organizations working with thousands of the world's most influential businesses and investors to fight climate change, summarized the main reasons for companies to implement low carbon strategies, taking into account both risk and opportunities (Fig. 7.2).

7.4 Barrier and Risk for Suppliers to Join the Carbon Neutral Chain

We already presented the main drivers for agri-food companies to move towards lower carbon emissions. Consequently, in this Section we will focus on the main barriers and risks for suppliers in implementing climate-friendly and carbon neutral strategies.

A study from Wreford et al. (2017) pointed out the main barriers farmers have to overcome to adapt their production towards more climate-friendly agricultural practices. The stated that these can occur both at farm level and at sector and policy level. The main barriers identified by Wreford et al. (2017) at farm level are:

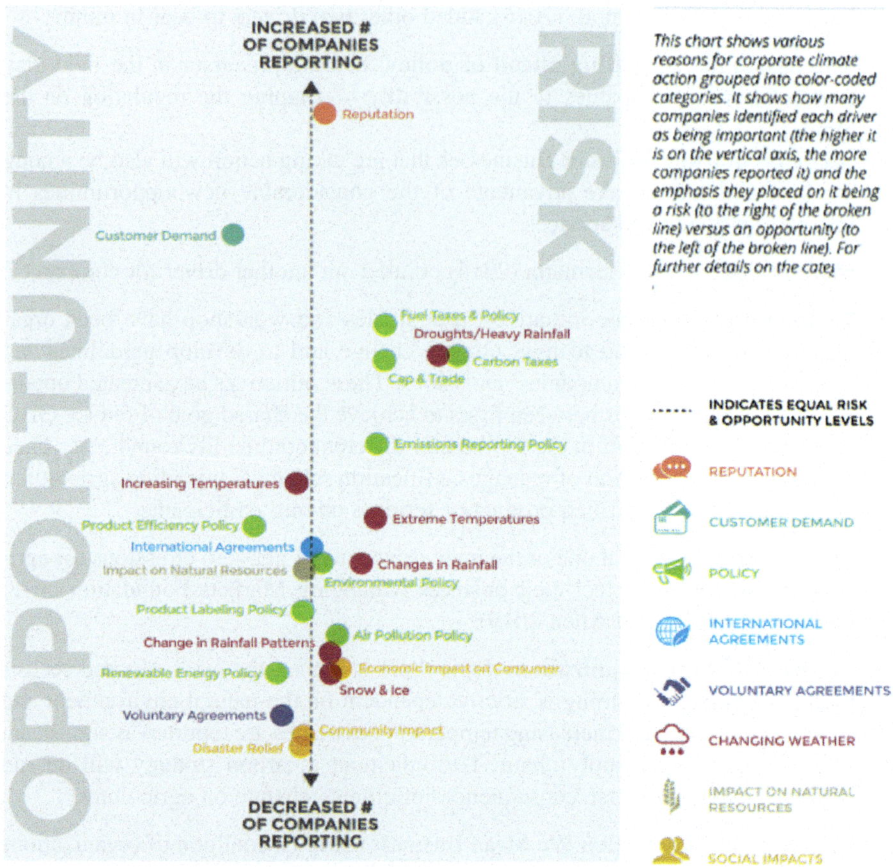

Fig. 7.2 Reasons for corporate climate actions
Source: We Mean Business (2015)

- **The non-possession of the land they cultivate**: Land tenure is a very important aspect in climate-friendly practices adoption. Farmers that rent or do not possess the land are less inclined to adopt long-term adaptation actions and to invest in them. (Claassen and Morehart 2011) stated that farmers that own their land may be more inclined to generate carbon offsets.
- **Absence of infrastructures**
- **Absence of farm succession plan and structure, and farmer age:** several studies demonstrated that the absence of farm succession plan is a barrier in planning and investing in a long term strategy for decarbonization, moreover, it has been demonstrated that big farms can more easily generate economies of scale and that young and well-educated farmers are more inclined to adopt sustainable agricultural practices (Diederen et al. 2003; Vanslembrouck et al. 2002).
- **Actual or perceived lack of financial benefits**
- **The cost of adopting new technologies or practices**

- **Hidden and transaction costs**: some studies argued that assessment of economic costs is often based on average value leading to an oversimplification. Moreover, there are some unobserved transaction cost related to learn and new agricultural techniques. Furthermore, as observed by Bellassen et al. (2015) carbon offset cost are the same for big and small farm companies presenting a significant barrier for smallholders. Particularly Smith et al. (2007) pointed out that brokerage cost in carbon market may represents a barrier for smallholders.
- **Limited access to credit**
- **Social and cultural factors:** farmers have a strong "cultural capital" and agricultural activity presents some aspect of tradition, identity and attachment to the land that may both obstacle the change and the implementation of climate-friendly practices.
- **Beliefs and experience of climate change have a modest impact on adoption**: as for consumers, to change their behavior farmers have to believe in climate change, understand what they can do to prevent and mitigate it and see that their actions have an actual impact.
- **Perceived long-time horizons, uncertainty about climate change effects and perceived relative low risk**
- **Farmers may attribute a low priority to climate mitigating measures.**

At policy and sector level Wreford et al. (2017) highlighted the following barriers:

- **Effect of practices on production:** at national scale, the possible decrease of production that could be caused by the implementation of climate-friendly practices is an important barrier
- **Lack of information and education awareness**
- **Industry co-operation:** the presence of competitive contract system that regulates the practices that can be implemented by farmers and the absence of an effective marketing chain may be important barriers to a climate-friendly agricultural production
- **The absence of and poorly designed climate policy**
- **Leakage:** Cohen and Vandenbergh (2012) suggested that some policymakers are concerned that "*GHG reductions in their home country will shift production to other countries with less stringent requirements, thereby reducing the impact of any climate legislation*". The carbon leakage is often used to argue the decision of nor reducing emission (Smith et al. 2007).
- **Reporting and administrative costs:** national level reporting of GHG emissions through the IPCC inventories and the fact that many mitigation practices are not captured by these inventories, as for example the benefits of carbon storage, represent a significant barrier.
- **Non-climate related agricultural policies:** Wreford et al. (2017) stated that "policies designed to support production, such as input subsidies, production support, subsidized insurance for marketable risks, and tax exemptions in challenging circumstances may distort the signals given to producers that they should be adopting climate friendly practices" (Table 7.2).

Table 7.2 Risks of carbon labelling for suppliers in developing countries

Risks of carbon labelling for suppliers in developing countries
Furthermore, Macgregor (2010) reviewed additional risks for carbon labelling implementation. He underlined how some suppliers in developing countries are concerned about the idea that these standards will become mandatory and that they can therefore jeopardize their ability to work in the agricultural trade market. Leaving uncertified agricultural products and small-scale producers from developing out of the global trade (Macgregor 2010).
Macgregor (2010) also highlighted that carbon labeling may result in trade issues with the label becoming a non-tariff yardstick. This aspect is pointed out also by Shi (2013) that argued that, due to spillover effects in developing countries, carbon labeling may represent a non-tariff barrier and act as a trade protection measure. Moreover, several authors stated that GHG emission calculations or Carbon Footprint (CF) calculation may be more difficult for developing countries, creating disadvantages (Brenton et al. 2009; Shi 2013). Also Plassmann et al. (2010) stressed that being designed by industrialized countries, CF calculation and PAS 2050, will not adequately represent production systems in developing countries. Moreover, the authors stated that both this aspect and the difficulties in estimating the GHG emissions related to land use change, will lead developing countries to suffer from trade restrictions.

7.5 Different Development Models for Carbon Neutrality

Currently, for the adoption of the carbon neutrality strategies have been different development models have been adopted. Mostly, the company's choice of adopting these strategies have been based on market drivers, they perceived benefits or market demand and they choose to conform to these market-driven requests. In this report, numerous case studies based on this type of adoption have been carried out. However, some other diffusion models can be found for Carbon neutrality development. Particularly, the government-driven model and sectoral legislation-drive ones. In the first, governments have taken actions to boost companies' interest in Carbon neutrality providing tools to developing it at home. This is the case of the French Label Bas Carbon. The second one is related to a specific sector that poses a limitation to the sector related GHG emissions and incentivize the compensation of these emissions. This is the case of the aviation industry. In the next sections, we will analyze these two models.

7.5.1 Government-Driven: The French Label Bas Carbone

The French Label Bas Carbone (French Carbon Standard) is a framework for voluntary carbon reduction project that was adopted by the French Government in November 2018. It provides a transparent framework for guaranteeing the integrity of carbon reduction projects (Directorate-General for Climate Action -European Commission 2019) (Fig. 7.3). Environmental integrity is ensured through the utilization of standardized methodologies in line with the overarching rules set in the regulation. To date, it includes approved methodologies for forestry (afforestation, coppicing, and restoration) and for agriculture (CARBON AGRI). Companies,

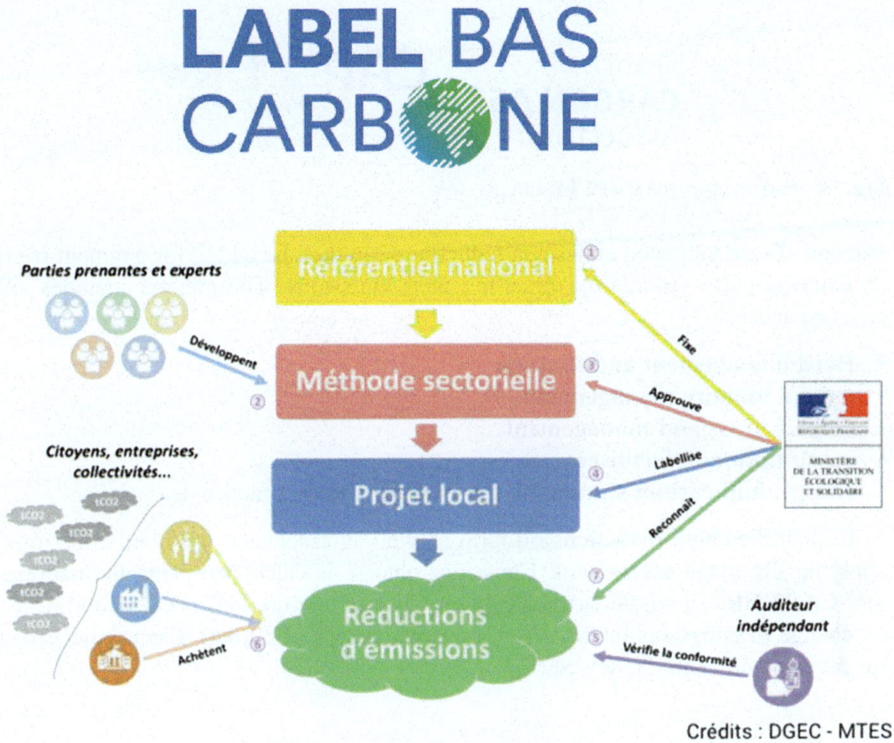

Fig. 7.3 Label Bas Carbone Logo and Development Model

public organizations or individuals that wish to compensate their emissions can voluntarily acquire the emission reductions determined thanks to these methodologies to offset their emissions. Individuals or sectors can propose methodologies, which the regulator must approve. These methodologies set guidelines for how to do the following: establish eligibility criteria, calculate baseline scenario and demonstrate additionality of the project, demonstrate environmental integrity (i.e. co-benefits), requirements on identifying and managing non-permanence risks, calculate emissions reductions relative to baseline, and MRV requirements and methods (Gabriella Cevallos et al. 2019).. Only projects that are additional will be approved (i.e. the carbon credits would shift the Net Present Value of the project from negative to positive and the project would not otherwise occur without credits). To date, it includes approved methodologies for forestry (afforestation, coppicing, and restoration) and for agriculture. For what concerns agriculture methodologies have been approved for: cattle and field crops (CARBON AGRI), sustainable management of hedges and plantation of orchards (Fig. 7.4).

CARBON AGRI provides a method for project developers (i.e. person/organization/company) to account for emissions reductions on **cattle (beef and dairy) or field crop farms** in France thanks to actions that **mitigate GHG** or **increase carbon**

Fig. 7.4 Carbon Agri and CAP2'ER Logos

storage. These validated emissions reductions can then be traded for payment from an external party voluntarily offsetting their emissions. The method includes six types of actions:

- **Herd management and feeding,**
- **animal manure management,**
- **crop & grassland management,**
- **consumption of fertilisers**
- **energy, and carbon storage** (in total 40 low carbon practices).

It quantifies both reductions on farm as well as associated upstream emissions, applying **life cycle assessment**. Emissions change is calculated using the national tool **CAP2'ER**®, a whole farm calculator. Change in emissions is calculated based on change in emissions intensity (i.e. kg GHG per kg of output). Each project runs for 5 years and can be renewed (Fig. 7.4).

7.5.2 Sectoral Legislation-Driven: The ICAO Model

In 2010, the International Civil Aviation Organization (ICAO) agreed on two goals regarding aviation emissions, namely the improvement of fuel efficiency by 2% annually through 2050 and a so-called Carbon Neutral Growth target to offset all pollution above 2020 levels. To reach this goal, a market-based measure, the Carbon Offset and Reduction Scheme for International Aviation (CORSIA) was adopted in 2016.

The scheme will enter into force in 2021, but participation will be voluntary until 2027. CORSIA requires individual airplane operators to compensate for their calculated share of emissions above their 2020 baselines using eligible emissions units and sustainable alternative fuels with demonstrably lower life-cycle emissions Depending on how the sector grows in the coming years, CORSIA may generate between 1.6 and 3.7 billion tons of demand for offsets between 2021 and 2035 (Ecosystem Marketplace, 2019).

The medium-term deal is expected to provide more than $40 billion in funding for climate projects and offset 2.6 billion tons of C02 emissions between 2021 and 2035 (Reuters, 2019). Airlines from 80 countries representing 77% of international air traffic have joined the deal's voluntary first phase between 2021 and 2026. It becomes mandatory from 2027 for states with large aviation industries (Fig. 7.5).

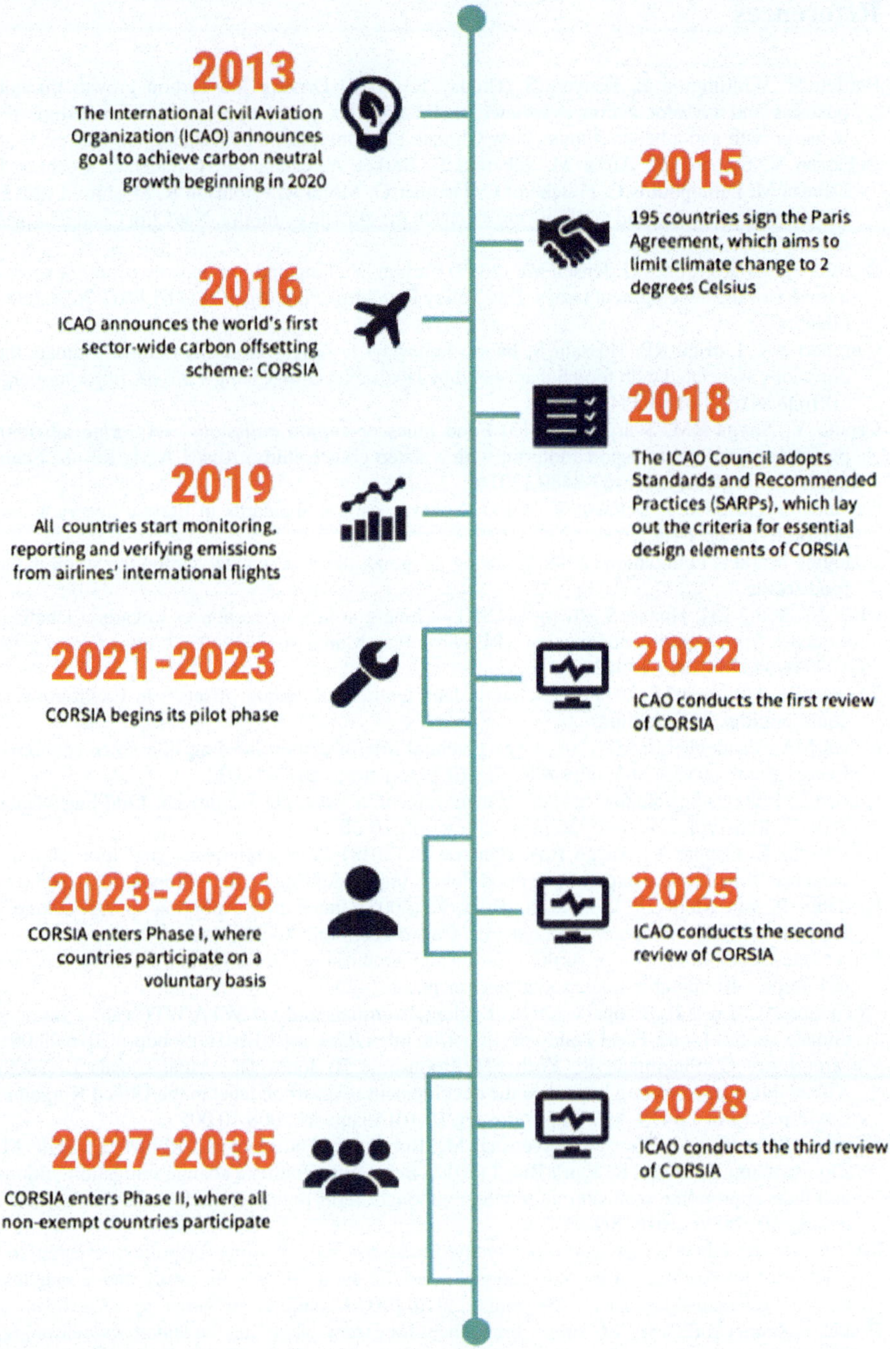

Fig. 7.5 The CORSIA timeline
Source: Forest Trends' Ecosystem Marketplace (2019)

References

Bartlett N, Whittington E, Reuvers S, Thoday K (2016) Driving low-carbon growth through business and investor action overview. In: Seizing the global opportunity: partnerships for better growth and a better climate. New Climate Economy, London/Washington, DC, pp 1–36

Bellassen V, Stephan N, Afriat M, Alberola E, Barker A, Chang JP, Chiquet C, Cochran I, Deheza M, Dimopoulos C, Foucherot C, Jacquier G, Morel R, Robinson R, Shishlov I (2015) Monitoring, reporting and verifying emissions in the climate economy. Nat Clim Chang. https://doi.org/10.1038/nclimate2544

Brenton P, Edwards-Jones G, Jensen MF (2009) Carbon labelling and low-income country exports: a review of the development issues. Dev Policy Rev. https://doi.org/10.1111/j.1467-7679.2009.00445.x

Camilleri AR, Larrick RP, Hossain S, Patino-Echeverri D (2019) Consumers underestimate the emissions associated with food but are aided by labels. Nat Clim Chang 9:53–58. https://doi.org/10.1038/s41558-018-0354-z

Caputo V, Nayga RM, Scarpa R (2013) Food miles or carbon emissions? Exploring labelling preference for food transport footprint with a stated choice study. Aust J Agric Resour Econ. https://doi.org/10.1111/1467-8489.12014

Cevallos G, Grimault J, Bellasse V (2019) Domestic carbon standards in Europe overview and perspectives, pp 1–44

Changing Markets Foundation (2018) Growing the good: the case for low-carbon transition in the food sector

Chen N, Zhang ZH, Huang S, Zheng L (2018) Chinese consumer responses to carbon labeling: evidence from experimental auctions. J Environ Plan Manag 61:2319–2337. https://doi.org/10.1080/09640568.2017.1394276

Claassen R, Morehart M (2011) Agricultural land tenure and carbon offsets*. In Environmental considerations in agriculture

Cohen MA, Vandenbergh MP (2012) The potential role of carbon labeling in a green economy. Energy Econ 34:S53–S63. https://doi.org/10.1016/j.eneco.2012.08.032

Corbett LM (2009) Sustainable operations management: a typological approach. J Ind Eng Manag 2:10–30. https://doi.org/10.3926/jiem.2009.v2n1.p10-30

De Marchi E, Caputo V, Nayga RM, Banterle A (2016) Time preferences and food choices: evidence from a choice experiment. Food Policy. https://doi.org/10.1016/j.foodpol.2016.05.004

Diederen P, van Meijl H, Wolters A, Bijak K (2003) Innovation Adoption in Agriculture : Innovators , Early Adopters and Laggards. Cah. d'économie Sociol. Rural

Directorate-General for Climate Action -European Commission (2019) Carbon Farming Schemes in Europe – Roundtable Background document, pp 1–23

Drichoutis AC, Lusk JL, Pappa V (2016) Elicitation formats and the WTA/WTP gap: a study of climate neutral foods. Food Policy 61:141–155. https://doi.org/10.1016/j.foodpol.2016.03.001

Edwards-Jones G, Plassmann K, York EH, Hounsome B, Jones DL, Milài Canals L (2009) Vulnerability of exporting nations to the development of a carbon label in the United Kingdom. Environ Sci Pol 12:479–490. https://doi.org/10.1016/j.envsci.2008.10.005

Faber J, Schroten A, Bles M, Sevenster M, Rohde C, Dütschke E, Köhler J, Gigli M, Zimmermann K, Soboh R, Van't Riet J (2012) Behavioural climate change mitigation options and their appropriate inclusion in quantitative longer term policy scenarios. CE Delft. https://doi.org/10.1007/s11010-009-0377-x

FAO (2010) Sustainable diets and biodiversity: directions and solutions for policy, research and action. In: Proceedings of the international scientific symposium biodiversity and sustainable diets united against hunger 3–5 November 2010. FAO Headquarters, Rome, pp 150–162

Feucht Y, Zander K (2018) Consumers' preferences for carbon labels and the underlying reasoning. A mixed methods approach in 6 European countries. J Clean Prod 178:740–748. https://doi.org/10.1016/j.jclepro.2017.12.236

Forest Trends' Ecosystem Marketplace (2019) Financing Emissions Reductions for the Future:State of the Voluntary Carbon Markets 2019

Gadema Z, Oglethorpe D (2011) The use and usefulness of carbon labelling food: a policy perspective from a survey of UK supermarket shoppers. Food Policy 36:815–822. https://doi.org/10.1016/j.foodpol.2011.08.001

Garnett T (2011) Where are the best opportunities for reducing greenhouse gas emissions in the food system (including the food chain)? Food Policy. https://doi.org/10.1016/j.foodpol.2010.10.010

Garnett T (2014) Changing what we eat: a call for research & action on widespread adoption of sustainable healthy eating. Food Clim. Res. Netw

Gassler B (2014) How green is your 'Grüner'? Millennial wine consumers' preferences and willingness-to-pay for eco-labeled wine. J Austrian Soc Agric Econ 24:131–140

Grunert KG, Hieke S, Wills J (2014) Sustainability labels on food products: consumer motivation, understanding and use. Food Policy 44:177–189. https://doi.org/10.1016/j.foodpol.2013.12.001

Hartikainen H, Roininen T, Katajajuuri JM, Pulkkinen H (2014) Finnish consumer perceptions of carbon footprints and carbon labelling of food products. J Clean Prod 73:285–293. https://doi.org/10.1016/j.jclepro.2013.09.018

Hertwich EG (2005) Life cycle approaches to sustainable consumption: a critical review. Environ Sci Technol. https://doi.org/10.1021/es0497375

InterTraderIreland (2011) Business & Carbon Measurement on the island of Ireland Acknowledgements

Iweala S, Spiller A, Meyerding S (2019) Buy good, feel good? The influence of the warm glow of giving on the evaluation of food items with ethical claims in the U.K. and Germany. J Clean Prod 215:315–328. https://doi.org/10.1016/j.jclepro.2018.12.266

Johnston JL, Fanzo JC, Cogill B (2014) Understanding sustainable diets: a descriptive analysis of the determinants and processes that influence diets and their impact on health, food security, and environmental sustainability. Adv Nutr. https://doi.org/10.3945/an.113.005553

Kima H, Houseb LA, Kim TK (2016) Consumer perceptions of climate change and willingness to pay for mandatory implementation of low carbon labels: the case of South Korea. Int Food Agribus Manag Rev 19:129–144. https://doi.org/10.22434/IFAMR2015.0095

Kolk A, Pinkse J (2005) Business responses to climate change: identifying emergent strategies. Calif Manag Rev 47:5–20

Leire C, Thidell Å (2005) Product-related environmental information to guide consumer purchases – a review and analysis of research on perceptions, understanding and use among Nordic consumers. J Clean Prod. https://doi.org/10.1016/j.jclepro.2004.12.004

Lombardi GV, Berni R, Rocchi B (2017) Environmental friendly food. Choice experiment to assess consumer's attitude toward "climate neutral" milk: the role of communication. J Clean Prod 142:257–262. https://doi.org/10.1016/j.jclepro.2016.05.125

Lóránt A, Allen B (2019) Net-zero agriculture in 2050: how to get there?

Macgregor BJ (2010) Carbon concerns: how standards and labelling initiatives carbon concerns: how standards and labelling, ICTSD–IPC platform on climate change. Agriculture and Trade, Geneva/Washington, DC

Mason P, Lang T (2017) Sustainable diets: how ecological nutrition can transform consumption and the food system. Routledge/Taylor and Francis Group. https://doi.org/10.4324/9781315802930

Murphy-bokern D, Kleemann L (2014) The role of corporate social responsibility in reducing greenhouse gas emissions from agriculture and food

Onozaka Y, McFadden DT (2011) Does local labeling complement or compete with other sustainable labels? A conjoint analysis of direct and joint values for fresh produce claim. Am J Agric Econ 93:689–702. https://doi.org/10.1093/ajae/aar005

Onozaka Y, Hu W, Thilmany DD (2016) Can eco-labels reduce carbon emissions? Market-wide analysis of carbon labeling and locally grown fresh apples. Renew Agric Food Syst. https://doi.org/10.1017/S174217051500006X

Pinkse J, Busch T (2013) The emergence of corporate carbon norms: strategic directions and managerial implications: Thunderbird. Int Bus Rev. https://doi.org/10.1002/tie

Plassmann K, Norton A, Attarzadeh N, Jensen MP, Brenton P, Edwards-Jones G (2010) Methodological complexities of product carbon footprinting: a sensitivity analysis of key variables in a developing country context. Environ Sci Pol. https://doi.org/10.1016/j.envsci.2010.03.013

Pomarici E, Vecchio R (2014) Millennial generation attitudes to sustainable wine: an exploratory study on Italian consumers. J Clean Prod 66:537–545. https://doi.org/10.1016/j.jclepro.2013.10.058

Sharp A, Wheeler M (2013) Reducing householders' grocery carbon emissions: carbon literacy and carbon label preferences. Australas Mark J 21:240–249. https://doi.org/10.1016/j.ausmj.2013.08.004

Shi X (2013) The spillover effects of carbon footprint labelling on less developed countries. Dev Policy Rev 31:239–254

Shuai CM, Ding LP, Zhang YK, Guo Q, Shuai J (2014) How consumers are willing to pay for low-carbon products? – results from a carbon-labeling scenario experiment in China. J Clean Prod. https://doi.org/10.1016/j.jclepro.2014.07.008

Sirieix L, Delanchy M, Remaud H, Zepeda L, Gurviez P (2013) Consumers' perceptions of individual and combined sustainable food labels: a UK pilot investigation. Int J Consum Stud 37:143–151. https://doi.org/10.1111/j.1470-6431.2012.01109.x

Smith P, Martino D, Cai Z, Gwary D, Janzen H, Kumar P, McCarl B, Ogle S, O'Mara F, Rice C, Scholes B, Sirotenko O, Howden M, McAllister T, Pan G, Romanenkov V, Schneider U, Towprayoon S (2007) Policy and technological constraints to implementation of greenhouse gas mitigation options in agriculture. Agric Ecosyst Environ. https://doi.org/10.1016/j.agee.2006.06.006

Sorensen H (2009) Inside the mind of the shopper: the science of retailing. Pearson Prentice Hall, Hoboken

Stokes AM, Turri A (2015) Consumer perceptions of carbon labeling in print advertising: hype or effective communication strategy? J Mark Commun 21:300–315. https://doi.org/10.1080/13527266.2012.762420

Thøgersen J, Nielsen KS (2016) A better carbon footprint label. J Clean Prod 125:86–94. https://doi.org/10.1016/j.jclepro.2016.03.098

Upham P, Bleda M (2009) Carbon labelling : public perceptions of the debate. Tyndall Cent. Manchester Sustain. Consum. Inst

Van Loo EJ, Caputo V, Nayga RM, Verbeke W (2014) Consumers' valuation of sustainability labels on meat. Food Policy 49:137–150. https://doi.org/10.1016/j.foodpol.2014.07.002

Vanclay JK, Shortiss J, Aulsebrook S, Gillespie AM, Howell BC, Johanni R, Maher MJ, Mitchell KM, Stewart MD, Yates J (2011) Customer response to carbon labelling of groceries. J Consum Policy 34:153–160. https://doi.org/10.1007/s10603-010-9140-7

Vanslembrouck I, Van Huylenbroeck G, Verbeke W (2002) Determinants of the willingness of Belgian farmers to participate in agri-environmental measures. J Agric Econ. https://doi.org/10.1111/j.1477-9552.2002.tb00034.x

We Mean Business (2015) The climate has changed

World Economic Forum (2018) Two Degrees of Transformation Businesses are coming together to lead on climate change . Will you join them?

Wreford A, Ignaciuk A, Gruère G (2017) Overcoming barriers to the adoption of climate-friendly practices in agriculture. OECD Food, Agric. Fish. Pap 101. https://doi.org/10.1787/97767de8-en

Chapter 8
Towards a Carbon Labelled World?

The ninth chapter explores the role of carbon labels in guiding consumers towards more environmentally friendly and low carbon choices and to inform and educate them about the environmental impact of the food's life cycle. Particularly, the chapter reviews the main carbon labels and consumers'perceptions about carbon neutral labels on the shelf and available to consumers. Moreover, the chapter also deals with carbon labels design and how a better design can help communicate environmental and carbon information. Additionally, the role of green marketing and communication has been explored, with a particular focus on how to communicate climate change and climate-related information to consumers. Finally, it explores the role of retailers and Green Public Procurements (GPP) in helping the spreading of these labels and boosts the interest of all the agri-food supply-chain in developing labels able to identify low carbon and carbon neutral products.

8.1 Consumers Perceptions and Use of Carbon Labelling

Food labels have been found in literature as an effective approach to guide consumers towards more environmentally friendly choices and to inform and educate them about the environmental impact of the food's life cycle (Camilleri et al. 2019). Consumers can be empowered by food labels helping them to make an informed choice (Emberger-Klein and Menrad 2018; Thøgersen et al. 2010). Labels and certification can be useful in transmitting information to consumers and can help companies to market and communicate their eco-friendly efforts (Sirieix et al. 2013). In this sense, the ISO 14020 series standards, helps companies to accurately communicate environmental information to consumers without being inaccurate and avoiding consumer confusion or the risk of greenwashing practices. In fact, these standards "*provides businesses with a globally recognized and credible set of international benchmarks against which they can prepare their environmental*

A. Acampora et al., *Carbon Neutrality in the Agri-food Sector*,
https://doi.org/10.1007/978-3-030-88048-4_8

labeling" (ISO Central Secretariat 2012). In the ISO 14020 series we can distinguish three types of labeling / ecological declarations:

- first type (ISO 14024)(ISO 2018): is a voluntary "third party" eco-label, for which there is a need for verification by an independent third-party organization, defined as the "competent body for the environmental labeling ", which certifies the compliance with specific predefined requirements. The predefined requirements to be met for obtaining the label are the "product environmental criteria", relating to an entire category of products, and which must be based on appropriate indicators deriving from the life cycle analysis for that specific category. The most authoritative example of labeling of the first type is the European Ecolabel regulation, an ecological quality mark born in 1992 with the adoption of the European Regulation n. 880/92 and updated with the new Regulation n. 66/2010.
- second type (ISO 14021:2016)(ISO 2016): labels and ecological declarations that contain environmental information or "self-declared environmental claims" (i.e. declarations, labels, environmental symbols on product packaging or advertising) declared by producers, importers or products distributors, without the intervention of an independent certification body (among which: "Recyclable", "Compostable", etc.). The standard provides however a series of constraints to be respected in terms of dissemination and information requirements.
- third type (ISO 14025:2010)(ISO 2010): ecological declarations containing information based on established parameters that contain a quantification of the environmental impacts associated with the life cycle of the product calculated through an LCA system. They are subject to an independent control and presented in a clear and comparable form. These include, for example, the "Environmental Product Declarations" or Environmental Product Declaration (EPD). This tool represents a communication method that guarantees objective, comparable and credible information relating to the environmental performance of products and services and aims to respond to the need to make the comparison between products of the same class easier.

The intangibility of environmental load of the food production and distribution processes can be overcome by using these labels to reduce consumers information asymmetry (Gold et al. 2017). In particular, Garnett et al. (2015) pointed out that, for this purpose, information on the environmental load of food items may not be enough to drive consumers behavioral change and food choice in the light of the attitude behavior gap. Nonetheless, to act as shopping aid towards low carbon foods choice, labels have to be familiar to consumers (Emberger-Klein and Menrad 2018). In fact, if they are not familiar or not completely understood, they can affect consumption choices in a negative way. To be effective, labels need to be noticed, understood (Leire and Thidell 2005), trusted, and value as a decision making tool (Sharp and Wheeler 2013).

However, from many studies emerges a lack of consumer awareness about food production and consumption environmental impact and about the impact that their food choices can play in mitigating GHG emissions (Camilleri et al. 2019; Leire and Thidell 2005; Lombardi et al. 2017). Additionally, several studies show that consumers are confused about the meaning and usefulness of carbon labels (Gadema and Oglethorpe 2011; Hartikainen et al. 2014), carbon neutral labels (Lombardi et al. 2017) and product carbon footprint meaning (Emberger-Klein and Menrad 2018).

Particularly, Gadema and Oglethorpe (2011) conducting a survey on UK supermarket shoppers (428 responses) found that 72% of respondents expressed a preference for carbon labels on food products. However, 81% either strongly agreed or agreed that understanding carbon footprint information and comparing carbon footprints was difficult and confusing. Only 14% of consumers stated that they knew their own personal carbon footprint and only a fifth of consumers thought a carbon footprint label on products would indicate better quality. Though, when asked whether consumers thought that carbon labels are useful to compare the environmental credentials of food products, 63% agreed.

In particular, Hartikainen et al. (2014) found that Finnish consumers experience several knowledge gaps that act as barriers in purchasing and using of low carbon food products:

- Knowledge gap 1: lack of understanding of the potential of making environmentally and climatically friendly choices by adjusting their food consumption habits;
- Knowledge gap 2: difficulties in understanding the factors that contribute the most to the environmental load of food;
- Knowledge gap 3: low level of understanding of the term "product carbon footprint".

This lack of understanding is particularly critical as many studies show that for carbon labels to be understood, and therefore effective in changing consumers purchase decision making, it is required for the consumers to have a base level of carbon knowledge and carbon labels literacy (Sharp and Wheeler 2013).

Grunert et al. (2014) explored European consumers' attitude (consumers' sample come from UK, France, Germany, Spain, Sweden, and Poland) towards different eco-labelled food products (i.e. chocolate, coffee, ice cream, breakfast cereal, ready meals and soft drink) using the motivation, ability and opportunity framework[1] proposed by Maclnnis and Jaworski (1989) to explain the role for labels effectiveness of information processing and consumer reaction to advertising information. This framework shows that advertising effects depend on: the consumer's opportunity to process the ad information, on the consumer's ability to interpret the ad information and make inferences from it, and on the consumer's motivation to engage in processing the information (Petty and Cacioppo 1986).

[1]The framework has been proposed by Maclnnis and Jaworski (1989) for explaining how consumers react to advertising information, noting that any advertising effects depend on the consumer's opportunity to process the ad information (related to the amount of exposure to the information), on the consumer's ability to interpret the ad information and make inferences from it, and fi- nally on the consumer's motivation to engage in processing the information, as the latter will determine the depth of processing and hence the type of memory trace that the ad message leaves (Petty and Cacioppo 1996).

However, Gadema and Oglethorpe (2011), in their study on UK groceries' shoppers, stated that whilst there is general confusion and misunderstanding about carbon labeling, this is not driven by an unwillingness to learn or an apathetic position. They suggest that consumers are willing to make choices around the carbon credentials of products but do not feel empowered to do so (Gadema and Oglethorpe 2011). In fact, several studies show that after being educated and informed about climate change seriousness, their ability to contribute to climate change mitigation and, about carbon labels meaning, consumers are more willing to change their consumption patterns towards low carbon choices (Emberger-Klein and Menrad 2018; Lombardi et al. 2017; Sirieix et al. 2013; Stokes and Turri 2015).

Moreover, the credibility of carbon labels credentials is a key factor to trigger consumers response towards these labels. In this sense a certifying and auditing system that assures consumers about the truthfulness of labels criteria and information provided to them is essential to influence their decision-making process (Sirieix et al. 2013; Teisl 2003). This aspect should be taken into consideration by the companies that are considering launching a private carbon-labeling program for food products (Sirieix et al. 2013). In their study on UK consumers' perception of eco-labelled fruits and vegetables, Sirieix et al. (2013) stated that food companies should carefully consider to have a: "good imagery, a clear message to promote and protect one's reputation, and the question of fit is of the utmost importance". As matter of fact, consumers tend to place more trust into products endorsed by a credible and familiar entity (Teisl et al. 2008).

Additionally, there is an issue that is worth some reflections. Companies have to be sure customers would appreciate their efforts towards low carbon food production, even if their products have high content of GHG emissions and with consistent environmental impacts. In fact, in this sense Thøgersen and Nielsen (2016) argued that is unlikely that companies would voluntarily have their product labeled with a red or a black label. The authors, indeed, advocated for a legal regulation or a government intervention to help labels proliferation. However, even if some studies revealed that carbon labels are more likely to impact consumers choices when used with products consider green by consumers (Beattie and Sale 2016), others suggests that carbon label, in this case the carbon neutral one, has more impact on consumer perceptions for an environmentally harmful product than for a less harmful product (Stokes and Turri 2015).

8.2 Carbon labels in Relation to Others Greening Themes and Labels

Over the last years there has been a proliferation of labels with environmental or social claims for food products (Gadema and Oglethorpe 2011). The exponential increase of these labels is mostly due to the growing environmental and social concerns of consumers (Sirieix et al. 2013). Consumers are increasingly paying

Fig. 8.1 Main Food Labels
Source: Authors' own elaboration

attention to the environmental and social impacts of their food choices and ask to be informed about the environmental and social burden of the foods that they are buying (Vecchio and Annunziata 2015). However, the recent eco-labels proliferation may also confuse consumers instead of informing them, especially if they are not well informed about environmental problems (Gadema and Oglethorpe 2011; Monier-Dilhan 2018; Zepeda et al. 2013). Sirieix et al. (2013) pointed out that the profusion of labels may be an opportunity for complementary between labels but can also add to the increasing competition among products in the consumers' choices (Padel et al. 2010).This is particularly true for carbon labels, indeed, recent research shows that climate change and climate-friendly products are less important to consumers than other sustainability attributes (Gadema and Oglethorpe 2011; Guenther et al. 2012; Hartikainen et al. 2014). In 2013 Ipsos and London Economics Consortium (2012) reviewed more than 900 European food labels schemes of which 174 concerned environmental management and 40 climate change. However, in this study environmental management and climate change scheme are those with the lowest level of awareness among European consumers (Fig. 8.1).

Scholars generally agree that the attributes that dominate consumers food choices still remain: quality, nutrition and price (Gadema and Oglethorpe 2011; Sharp 2010; Sirieix et al. 2013). However sustainability claims appeal to consumers is well supported in literature (Grunert and Aachmann 2016; Janßen and Langen 2017; Van Loo et al. 2014).

Generally, scholars have demonstrated that consumers are more frequently willing to pay higher price premiums for local and organic production and animal welfare indications than for carbon labels (De Marchi et al., 2016; Onozaka and McFadden 2011; Van Loo et al. 2014).

Feucht and Zander (2018) study shows that local production is particularly important for Europeam consumers (consumers' sample come from France, Germany, Italy, Norway, Spain and United Kingdom) as they willing to pay higher price for locally produced products than for climate indications and organic production. Also in Lampert et al. (2017) study the origin of the product is the most important information for different asparagus choice for German consumers. However, they found that to a smaller group of consumers who are already aware of climate change problem and carbon labels the information on the carbon footprint of vegetables can create additional value. Emberger-Klein and Menrad (2018) study results shows that origin and production method are the most important aspect considered by German consumers in buying tomatoes, while carbon labels were of minor importance in the decision-making process. Several studies found that environmental concern is a primary consumers' driver in buying organic and local foods (Bauer et al. 2013; Nie and Zepeda 2011). Indeed, consumers attach to organic (for less pesticides and fertilizers) and local (for less food miles and contribution to the local economy) products environmental protection attributes (Adams and Salois 2010; Feucht and Zander 2018). Feucht and Zander (2018), stated that consumers associate more personal and ethical benefits from "organic" and "local" than from "climate-friendly" attributes. The authors suggested that consumers tend to consider both "organic" and "local" under the umbrella of "eco-friendly behavior"(Feucht and Zander 2018). Onozaka and McFadden (2011) pointed out that this may create a "substitution effect" for consumers in considering other sustainability claim (such as carbon information), since with the purchase of an organic or local product they are already addressing the issue of sustainability.

Additionally, in Gadema and Oglethorpe (2011) free range and fair trade has been found the most important issue for UK consumers than carbon emission, being evaluated as the second lowest attribute for importance in the choice of food items. Also Vecchio and Annunziata (2015) findings support the Italian consumers appreciation of the Fair Trade label. In their study they found that young individuals are more willing to pay for Fair Trade labelled chocolate than Rain Forest Alliance label and Carbon labeled chocolate. Additionally, in their study about Australian consumers preference for sustainability labels in groceries, Sharp and Wheeler (2013) revealed that durables' labelling (such as Energy and Water Ratings) had the most claimed influence in determining consumers choice, but that the Australian made, Fair Trade, Marine Stewardship Council and Organic labels, achieved respectively 63%, 51%, 44% and 39% of claimed influence. Additionally, McGuire and Beattie (2016) simulated a shopping environment to test differences in UK consumers behavior when shopping alone or with friends. Their results showed that when consumers are with their friends are more likely to buy organic or ecological products. In their study, however, low-carbon products are not chosen by consumers who buy with friends. This could mean that these labels are not well recognized or "obvious enough" to allow social signaling, or they are worried that their friends will not be able to recognize and appreciate carbon label added value. Considering as above stated, Sirieix et al. (2013) investigated the possibility of combining different sustainability labels. However, their finding leads to exclude this possibility because the participants did not like any of the labels combinations or they felt the combinations were contradictory. Also Lombardi et al. (2017) pointed out a possible substitution effect between different labels. In their study they found that after informing consumers about climate change issues, the utility of carbon free milk increased, whereas the utility of organic milk decreased. Similarly, Onozaka and McFadden (2011), exploring american consumers' willingness to pay (WTP) for eco-friendly apples and tomatoes, found that consumers discounted more severely carbon-intensive local products are than those sourced from other locations. Contrarily, Akaichi et al. (2017), in their study on american consumers'WTP for eco-certified rice, found that information on GHG emissions and localy produced and GHG emission and food miles complement each other, inceasing consumers WTP.

8.3 Carbon Labels: Does Design Matter?

Given consumers lack knowledge of the problems related to climate change and the limited use by carbon labels or carbon neutrality labels, researchers hypothesized that this is due to the lack of information and education of consumers. The labels, as previously mentioned, represent the tool through which educate and inform the final consumers (Sharp and Wheeler 2013). Consumers do not feel empowered enough to make choices about carbon labels because they are disoriented about the true meaning of these labels (Gadema and Oglethorpe 2011). In this sense labels design matters (Meyerding et al. 2019; Thøgersen and Nielsen 2016). From the literature emerged the need to design carbon labels which are easily understood by consumers and that can guide consumers through more climate friendly food choices (Upham et al. 2011; Upham and Bleda 2009). The ability of carbon labels to help mitigating climate change issues ultimately depends on consumer response to labelling (Elofsson et al. 2016). Scholars suggested that eco-label success depends on several factors such as: label design, the type of product labelled, and how and where the product is marketed (Elofsson et al. 2016; Hallstein and Villas-Boas 2013; Onozaka et al. 2010; Vlaeminck et al. 2014). For what concerns label design, scholars generally agree that the traffic light color ranking (red, yellow, green) is the most successful in guiding consumers towards more low carbon choices (Camilleri et al. 2019; Emberger-Klein and Menrad 2018; Meyerding et al. 2019; Sharp and Wheeler 2013; Thøgersen and Nielsen 2016) (Fig. 8.2). This color combination is familiar to consumer, easy to understand and helps them to make comparison between several products. The combination is also preferred to the black-yellow-green one (Sharp and Wheeler 2013) (Fig. 8.3).

The color black was initially chosen since it has been seen as the 'carbon color' used in marketing and media claims (Sharp and Wheeler 2013; Vanclay et al. 2011) and by Carbon Trust certification scheme. However, researchers pointed out that the black footprint as adopted by Carbon Trust and by the retailer Tesco has been found difficult to understand by consumers (Gadema and Oglethorpe 2011; Hartikainen

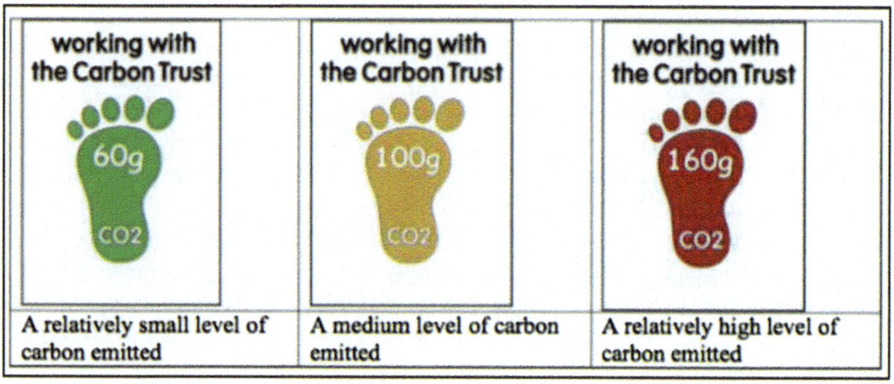

Fig. 8.2 The green-yellow-red combination
Source: Sharp and Wheeler (2013)

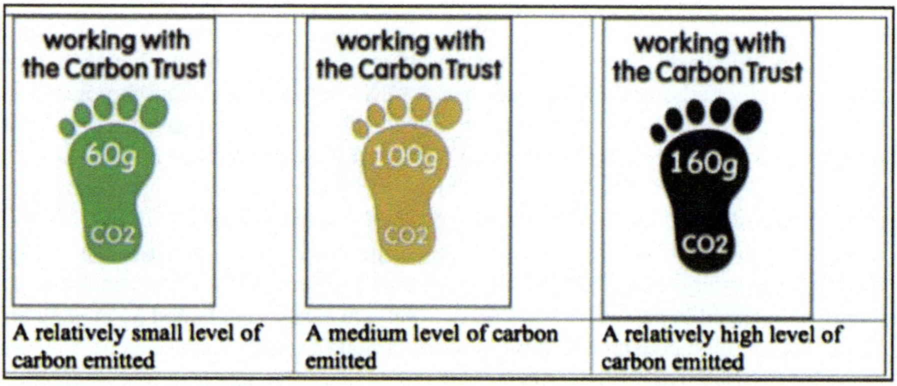

Fig. 8.3 The green-yellow-black combination
Source: Sharp and Wheeler (2013)

et al. 2014; Hornibrook et al. 2013), also leading to the abandonment of the Tesco efforts due to insufficient uptake by consumers (Meyerding et al. 2019). The low level of appeal to consumers of the Carbon Trust label is also due to the fact that consumers are not able to classify an abstract number as a continuous variable of greenhouse gas emissions (Camilleri et al. 2019; Meyerding et al. 2019; Sharp and Wheeler 2013) (Fig. 8.4).

To overcome this issue, Camilleri et al. (2019) suggested to include reference values allowing consumers to compare GHG emissions and to put information in context, proposing to express GHG emissions in terms of a familiar unit, as for example equivalent light-bulb minutes (Camilleri et al. 2019). Other scholars, instead, suggested to use a 5 layers scale helping consumers to compare between a specific food product category (Feucht and Zander 2018; Hartikainen et al. 2014; Kortelainen et al. 2016; Meyerding et al. 2019)(Fig. 8.5).

Fig. 8.4 Carbon trust label

Fig. 8.5 Carbon scale
Source: Hartikainen et al. (2014)

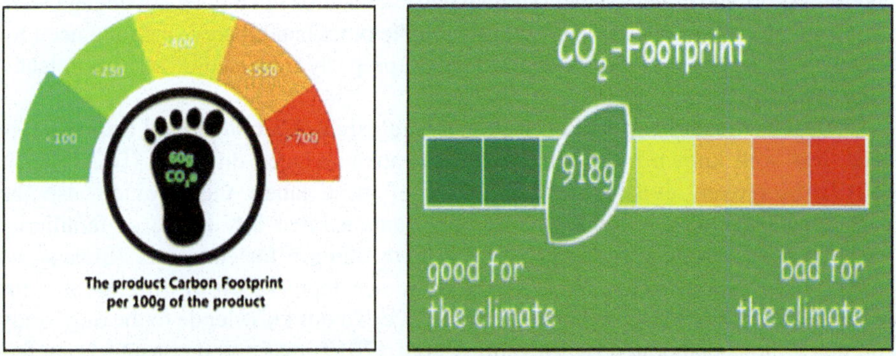

Fig. 8.6 Carbon scale and GHG emissions
Source: Feucht and Zander (2018), Meyerding et al. (2019)

However, Feucht and Zander (2018) argued that some consumers still prefer an absolute number of GHG emissions that allows them to make comparisons between different food products categories suggesting a combination of scale and an absolute number of CO^2-equivalents (Fig. 8.6).

Furthermore, Zhou et al. (2019) demonstrated that placing the carbon label on the right in the visual field of packaging with an horizontal allocation of space increases consumers'purchase intention toward a green product (Fig. 8.7).

Fig. 8.7 Label on the right visual field with horizontal allocation
Source: Zhou et al. (2019)

However, marketers are faced with a dilemma, if on the one hand consumers prefer a short and concise label (Upham and Bleda 2009) that does not engage them in a long process of reading and scrutiny, on the other hand there is a strong need for detailed information at the consumer level to properly understand the carbon labelling (Meyerding et al. 2019).

This is particularly critical for carbon neutral labels. Carbon neutral claims are, in fact, less well known by consumers, due to their reduced diffusion (Iweala et al. 2019). Moreover, due to the complexity of these labels that provide also the offsetting possibility to which consumers demonstrated low levels of familiarity (Peattie et al. 2009), they could be even more obscure for consumers(Stokes and Turri 2015). Additionally, literature on consumer appreciation of different designs and understanding for these labels is very scarce. To our knowlegde to the only study on this topic is elaborated by Meyerding et al. (2019) that tested consumers preference for three different labels (Fig. 8.8).

Results from this study reported positive consumers appreciation values for the "carbon neutral label" and the "carbon reduction label", even if they are relatively low, unlike the "the Carbon Trust label" that shows a negative part-worth utility. This result is in line with Hartikainen et al. (2014) findings pointing out the modest appreciation of the carbon neutrality label from the participants of their study. Concluding, labels strongly compete for consumers attention (Emberger-Klein and Menrad 2018) and for this reason a great design can help labels to stand out on the shelf and eye-catch consumers preference (Sharp and Wheeler 2013).

Fig. 8.8 Carbon trust, carbon label and carbon reduction label
Source: Meyerding et al. (2019)

8.4 How to Communicate Climate change and Products Carbon Information

Communicate climate change and carbon information to consumers is very challenging as levels of concern, carbon literacy and attitudes towards the actions they can undertake varies considerably among different segment of the market. In this sense marketing managers play a pivotal role in choosing how to engage and targeting consumers based on their predisposition towards environmental issues. In order to alter consumers behavior it is fundamental that they perceive the problem (e.g. climate change) as real, they need to understand how they can contribute, they have to feel as their contribution is relevant and that this contribution can make a difference (Peattie et al. 2009). Climate change is often perceived by consumers as an issue very distant from everyday problems and they feel emotionally distant from the problematic as if climate change does not affect them but "future generations" or people distant from them (Anable et al. 2006).

Peattie et al. (2009) in their article "*Climate change: A social and commercial marketing communications challenge*" pointed out some consideration that companies and marketers should bear in mind when planning their green marketing and communication strategy:

- climate change should be presented as issue "personally" relevant (e.g. prefer expression as "your planet" "your home");
- negative "guilt-based messages" are less effective; green communication that aims to play on consumers sense of guilt or fear it is not much supported in literature; this does not mean that it is not effective but sometimes use this strategy may alienate the consumer instead of engage him;
- low-carbon behavior should be associated with a "positive self-images" and less stressful lifestyle and not as sacrifice or deprivation;
- address climate change as "common enemy" to fight in a way that can bring people together and inspire people to act;
- focus on simple and effective actions that consumers can undertake here and now;

		Sample	House-holder[1]			Sample	House-holder[1]
n		413		House-hold size (%)	1 person	13.3	26.3
Gender (%)	female	69.0	67.3		2 persons	39.5	38.1
	male	31.0	32.7		3+ persons	47.2	35.6
Age (%)	16-29 years	24.7	11.2	Education[2] (%)	Low	3.4	11.9
	30-49 years	31.2	34.5		Medium	65.9	70.8
	50+ years	44.1	54.3		High	30.8	17.3

[1]Individuals 16+ years from BW being mainly or occasionally responsible for the shopping of the household and which bought food from EDEKA during the last months.

[2]Low: no/not yet a school leaving certificate, Certificate of Secondary Education (CES) without apprenticeship, Medium: CES with apprenticeship, secondary school, higher education entrance qualification High: higher education.

Fig. 8.9 Socio-demographic characteristics of the sample
Source: Klein and Menrad (2016)

- connect climate change issues on the land and the territory, appealing to consumer attachments to their homes;
- foster people sense of community and solidarity call them to act not just as individuals but as a community;
- prefer an interactive and entertaining as well as informative style of communication.

However, the major challenge that companies, marketers and governments have to deal with is the consumers' "greenwashing" skepticism. Greenwashing refers to: "*intentionally misleading or deceiving consumers with false claims about a firm's environmental practices and impact*" (Wang 2017). When consumers feel that a company claim is "greenwashed", they lose trust in that brand compromising brad equity. It is therefore, crucial that companies' efforts are seen by consumers as a genuine commitment to sustainability and climate change mitigation.

Sirieix et al. (2013) suggest that companies should prevent the greenwashing effect and address consumers concerns about it head-on with facts. One way to do this is by engaging consumers by telling a story about company sustainability commitment and explaining to them why sustainability and low carbon practices are important for company success. Additionally, eco-label can contrast the "distortive effect" caused by greenwashing and the consequent mistrust of consumers as they provide accepted and recognized certification schemes that can assure to consumers credible and clear information (Testa et al. 2015).

Moreover, Klein and Menrad (2016) explored the German consumers preferences for provision of information and supporting-instruments for climate consciousness choices. They used a quantitative survey with a sample of 413 persons primarily responsible for the household's grocery shopping. Figure 8.9 summarized the main socio-demographic characteristic of the sample.

Their quantitative study results highlighted that the 67% of their respondents found a climate label useful for the low-carbon food decisions, followed by information provided directly at the point-of-purchase (57%), and a joint label for positive attributes of food, and television reports on nutrition and climate. The instruments that they found less preferred by consumers are QR-Codes on the

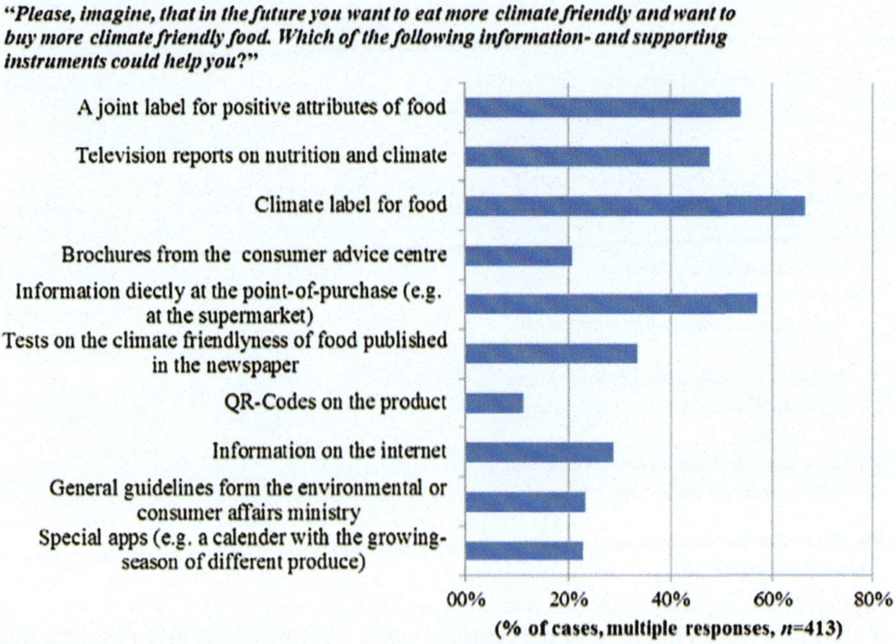

Fig. 8.10 Preferred information and supporting instruments
Source: Klein and Menrad (2016)

product (11%), brochures from the consumer advice center (21%), general guidelines from the environmental and consumer affairs ministry (23%) as well as special apps (23%) (Fig. 8.10).

However, differences between different socio-demographic groups in preferred information and supporting instruments have been found. Particularly, the most preferred supporting measure has been found the climate label for both men and women, and the QR-Code has been found as the least preferred one. The joint label is preferred by men, instead, brochures from the consumer advice center and newspaper reports are preferred by women. In general, women have a higher preference for general guidelines and apps. All age groups evaluate the climate label as the most suitable supporting instrument. However, young people (<29 years) like the idea of apps, while the older age groups reject this. Additionally, older people (50+ years) find that information on the internet is not so helpful for making more climate-friendly food choices, while both younger age groups (<29 years and < 49 years) like this idea. In contrast, the middle agers (<49 years) and older respondents (50+ years) have a higher preference for information from the consumer advice center.

Nonetheless, investigating consumers perception of eco-labels in the food sector Grunert (2011) founded six main possible barriers to their behavioral change represented in Fig. 8.11 below:

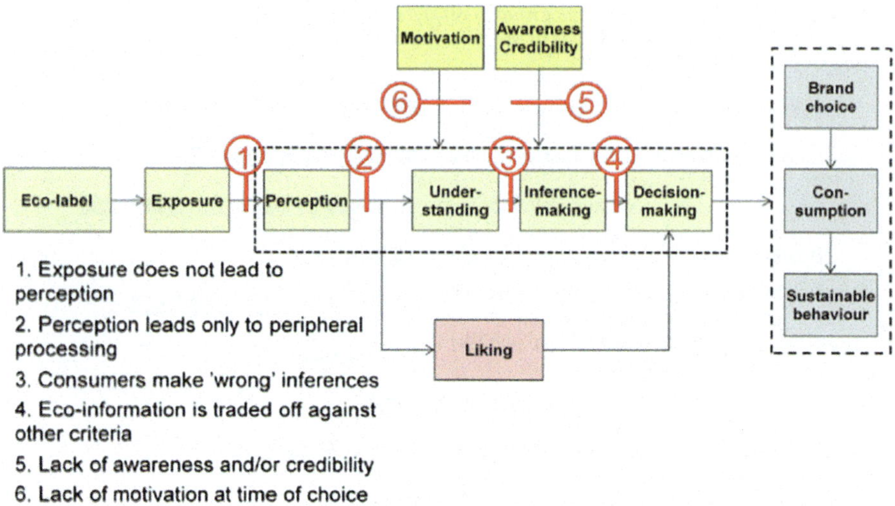

Fig. 8.11 Six barriers on food eco-labels effectiveness
Source: Grunert (2011)

1. **Exposure does not lead to perception:** labels are not noticed by consumers due to the information-overloaded environment and short time in which they shop and because they make food choices on a habitual basis.
2. **Perception leads only to peripheral processing:** according to the author, peripheral processing is a shallow form of processing that lead consumers to note the label, appreciate it, but don't make an effort to understand the true meaning of the label
3. **Consumers make 'wrong' inferences:** this type of barrier concern the difficulties of consumers to understand eco-label meaning. Contrary to the point above consumers try to understand what the label means but the fail to comprehend the label due to wrong inferences they make on it.
4. **Eco-information is traded off against other criteria:** when choosing a food products consumer have to decide among different criteria, sustainability criteria is often trade-off with other criteria as price and quality.
5. **Lack of awareness and/or credibility.** Consumers labels and sustainability literacy is fundamental in altering their food choices. Consumers have to feel empowered to make sustainability choices and they have to consider the eco-label credible
6. **Lack of motivation at time of choice.** The author explores the attitude-behavior gap stating that a time of choice even if they have a positive attitude towards the eco-label, this can be not strong enough to affect consumer choice and their behavior.

To elaborate a successful communication strategy, marketing managers should therefore address these barriers by assisting consumers in making more climate-friendly choices and helping them to change their behavioral patterns.

These barriers are even more critical when considering consumers perception of carbon neutral products. Especially for what concern the practice of carbon offsetting, this has often been seen by consumers as a greenwashing practice and generally perceived with skepticism (Bunning et al. 2013). This criticism regarding carbon offset has led to the diffusion of the term "carbon indulgence". Offset have been seen as a way for polluting companies to wash their sins by paying an indulgence, the offset credit, (as in indulgences of the Middle Ages) and continue to pollute (Nerlich and Koteyko 2009). Offset practices are also been criticized as they have been seen as a way for companies to pay to pollute rather than adjusting their production to more eco-efficient solutions and to reduce their emissions (Hyams and Fawcett 2013).

Additionally, Birkenberg and Birner (2018) pointed out that further reasons for this negative image are:

- a confusing set of related terms, such as zero-carbon, carbon-neutral, carbon- free and climate-neutral;
- lack of transparency on how carbon reductions were achieved and where the carbon credits came from;
- a lack of comparability due to missing standardized methodologies and differences in functional units.

Moreover, in terms of labels design and information provision, Schaefer and Blanke (2014) highlighted that often carbon neutral labels do not communicate to what extent the GHG emissions from the LCA study were reduced and the offset amounts.

For these reasons the UK's Department of Energy and Climate Change (DECC) has developed a series of guidelines to communicate carbon neutrality, to prevent "greenwashing" and to also provide greater clarity around the labels (Department of Energy & Climate Change 2009):

- Carbon neutral statement should provide clear information on the emissions measured, the reductions made, and the offsets purchased;
- Carbon neutrality claims should always be linked to period of time so that claim is understandable and transparent. This can also assure consumers that carbon neutrality assertions are reviewed and updated. Moreover, carbon neutral statements should be reviewed yearly;
- Carbon neutrality claims have to be accurate, truthful, relevant and avoiding the use of ambiguous terminology;
- Companies have to be clear about the scope of emissions measured. It should be clear that the term carbon neutral only applies to the specific certified scope;
- Is preferably not made a carbon neutral claim consisting only of calculating emissions and offsetting;
- It is recommended that companies use verified labels or claim for their carbon neutral statements.

8.5 The Role of retailers and Green Public Procurements (GPP) in Boosting Carbon Reduction and Carbon Neutral Labels

Even if carbon labeling has been created to boost consumers awareness on GHG emission and to influence their food choices, these labels also exerts pressure on manufacturers, retailers, and companies (Liu et al. 2016).

In particular, for what concern retailers, Emberger-Klein and Menrad (2018) suggested that given the privileged location between consumers and producers, they can strongly influence consumers in choosing low carbon products and reducing GHG emissions associated with food consumption (Ekelund et al. 2014; Styles et al. 2012). In fact, several scholars found that having carbon information at point of sale (POS), help consumers to make informed choices and increases the probability of engage them in choosing carbon labeled products (Elofsson et al. 2016; Leire and Thidell 2005; Sharp and Wheeler 2013). Additionally, retailers could adjust the criteria by which they choose their suppliers by requiring carbon labeling or by removing from their shelfs products with an high carbon footprint (Feucht and Zander 2018).

However, as stated by Ekelund et al. (2014) retailers are at a standstill in this field. The only retailers that has tried to implement this company policy, Tesco, the largest groceries retail in the UK, abandoned this project because it was not successful among consumers (Gadema and Oglethorpe 2011). Ekelund et al. (2014) explained this retailers immobility ascribing it to retailer's unwillingness to damage the meat segment (very impactful in terms of GHG emissions) or to encourage consumers to eat less food and to their difficulties or inability in communicating climate friendly products in the right way. These issues suggest that government intervention with specific policies to promote and enhance labels may be necessary (Gadema and Oglethorpe 2011). In this sense Green Public Procurements (GPP) may play a pivotal role in the climate friendly labels diffusion and success. GPP has been defined by the European Union as a: "*process whereby public authorities seek to procure goods, services and works with a reduced environmental impact throughout their life cycle when compared to goods, services and works with the same primary function that would otherwise be procured*" (Commission of the European Communities 2008). This process is often achieved through the definition of specific environmental criteria or standards in the call for tenders (Cerutti et al. 2016). Food products and catering services contracted for the education sector, the healthcare and welfare sector hospitals, the defense sector, the judicial sector and the government office canteens, representing a significant share of public sector procurement. Therefore, agri-food business may play a pivotal role in green public procurement application (Neto and Gama Caldas 2018; Smith et al. 2016). Soldi (2018) reviewed some European experiences dealing with the implementation of GPP and Sustainable Public Procurement (SPP) of public catering, showing the great contribution this type of instrument could bring to the development of a more sustainable food system (Table 8.1).

Table 8.1 The Italian GPP Action Plan

The Italian GPP action plan
In Italy, for example, the GPP National Action Plan, initially released in 2008, has been integrated in 2011 with the definition of Minimum Environmental Criteria (CAM) for catering services and food supply (Italian Ministry for the Environment Land and Sea 2011). In the Italian Minimum Environmental Criteria, beside criteria as food from organic production, food from integrated production, food quality, freshness and healthiness as well as regional food specificities (Filippini et al. 2018), also the carbon footprint has been inserted as a "further rewarding criterion" for the catering service and the food supply for which: "*It is possible to assign scores to the bid with the lowest number of GHG emission associated expressed in terms of CO^2eq, related to the entire life cycle of the service covered by the contract*"(Italian Ministry for the Environment Land and Sea 2011). However, according to a report by Ecosistemi (2012), published on the website of the Italian Ministry for the Environment Land and Sea, only in the 9% of the cases analyzed by the report, a rewarding criterion is inserted for those catering services who contemplate the Carbon Footprint.

Also, Neto and Gama Caldas (2018) reviewed environmental criteria for GPP schemes of EU Member States for public purchasing of food products and catering services. They found that the most used environmental criteria in GPP are: organic production, seasonal and fresh produce, staff training, transportation and packaging, menu planning, waste management (including food waste), marine and aquaculture products and animal welfare (Neto and Gama Caldas 2018). However, carbon emissions are never cited specifically. In fact, Cerutti et al., (2016) advocated that the European Union in GPP in food sector takes more in consideration sustainability practices than threshold values (e.g. GHG emission) in specific environmental impact categories that do not allow quantification of the actual environmental savings. This is supported also by Tsai (2017) and Alvarez and Rubicon (2015) that suggested the inclusion of carbon footprint in GPP and green-mark products. To overcome this issue, around 2010 Low carbon procurement (LCP) emerged as a possible direction and a subset of GPP (Cheng et al. 2018). LCP has been defined by Correia et al. (2013) as a: "*process whereby organizations seek to procure goods, services, works and utilities with a reduced carbon footprint throughout their life cycle and/or leading to the reduction of the overall organizational carbon footprint when considering its direct and* indirect emissions". Correia et al. (2013) highlighted that LCP is an emerging concept that will have an increasing role in GPP. Nonetheless, to be effective LCP will need more carbon literacy and understanding of carbon management concepts, methodology and complexity by procurers and policymakers (Correia et al. 2013).

In recent years, green procurement has gained popularity also among private sector companies that have started to integrate environmental aspects into decision making along the entire supply chain (Appolloni et al. 2014; Cheng et al. 2018). In this sense, companies to greening their supply chain are choosing more environmental-friendly suppliers and are committing their self to purchase more green products (Lee 2008).

In particular, Abdallah et al. (2012) tested the effectiveness of green procurement in model in which the company decides on its suppliers based on the trade-off between suppliers' emissions and the cost of their components. This opens up different opportunities for companies with carbon labelled products whether they decide to operate in the public sector or in the private sector (Mungkung et al. 2018).

References

Abdallah T, Farhat A, Diabat A, Kennedy S (2012) Green supply chains with carbon trading and environmental sourcing: formulation and life cycle assessment. Appl Math Model. https://doi.org/10.1016/j.apm.2011.11.056

Adams DC, Salois MJ (2010) Local versus organic: a turn in consumer preferences and willingness-to-pay. Renew Agric Food Syst. https://doi.org/10.1017/S1742170510000219

Akaichi F, Nayga RM, Nalley LL (2017) Are there trade-offs in valuation with respect to greenhouse gas emissions, origin and food miles attributes? Eur Rev Agric Econ. https://doi.org/10.1093/erae/jbw008

Alvarez S, Rubicon A (2015) Carbon footprint in green public Procurement: A case study in the services sector. J Clean Prod. https://doi.org/10.1016/j.jclepro.2015.01.048

Anable J, Lane B, Kelay T (2006) An evidence base review of public attitudes to climate change and transport behaviour. Final Rep. to Dep. Transp. July, London

Appolloni A, Sun H, Jia F, Li X (2014) Green procurement in the private sector: a state of the art review between 1996 and 2013. J Clean Prod. https://doi.org/10.1016/j.jclepro.2014.08.106

Bauer HH, Heinrich D, Schäfer DB (2013) The effects of organic labels on global, local, and private brands. More hype than substance? J Bus Res. https://doi.org/10.1016/j.jbusres.2011.12.028

Beattie G, Sale L (2016) Explicit and implicit attitudes to low and high carbon footprint products. Int J Environ Cult Econ Soc Sustain Annu Rev. https://doi.org/10.18848/1832-2077/cgp/v05i04/54652

Birkenberg A, Birner R (2018) The world's first carbon neutral coffee: lessons on certification and innovation from a pioneer case in Costa Rica. J Clean Prod 189:485–501. https://doi.org/10.1016/j.jclepro.2018.03.226

Bunning J, Beattie C, Rauland V, Newman P (2013) Low-carbon sustainable precincts: an Australian perspective. Sustain. https://doi.org/10.3390/su5062305

Camilleri AR, Larrick RP, Hossain S, Patino-Echeverri D (2019) Consumers underestimate the emissions associated with food but are aided by labels. Nat Clim Chang 9:53–58. https://doi.org/10.1038/s41558-018-0354-z

Cerutti AK, Contu S, Ardente F, Donno D, Beccaro GL (2016) Carbon footprint in green public procurement: policy evaluation from a case study in the food sector. Food Policy 58:82–93. https://doi.org/10.1016/j.foodpol.2015.12.001

Cheng W, Appolloni A, D'Amato A, Zhu Q (2018) Green public procurement, missing concepts and future trends – a critical review. J Clean Prod 176:770–784. https://doi.org/10.1016/j.jclepro.2017.12.027

Commission of the European Communities (2008) Public procurement for a better environment, Communication from the Commission to the European Parliament, the Council, the European economic and social Committee and the Committee of the Regions

Ipsos and London Economics Consortium (2012) Consumer market Study on the functioning of voluntary food labelling schemes for consumers in the European Union EAHC/FWC/2012 86 04

Correia F, Howard M, Hawkins B, Pye A, Lamming R (2013) Low carbon procurement: an emerging agenda. J Purch Supply Manag 19:58–64. https://doi.org/10.1016/j.pursup.2012.11.004

Department of Energy & Climate Change (2009) Guidance on Carbon Neutrality [WWW Document]

Ecosistemi (2012) Indagine Ristorazione 2012: Stato di applicazione dei Criteri Ambientali Minimi negli appalti pubblici di ristorazione collettiva
Ekelund L, Hunter E, Spendrup S, Tjärnemo H (2014) Communicating GHG mitigating food consumption in-store. Br Food J. https://doi.org/10.1108/bfj-03-2013-0075
Elofsson K, Bengtsson N, Matsdotter E, Arntyr J (2016) The impact of climate information on milk demand: evidence from a field experiment. Food Policy. https://doi.org/10.1016/j.foodpol.2015.11.002
Emberger-Klein A, Menrad K (2018) The effect of information provision on supermarket consumers' use of and preferences for carbon labels in Germany. J Clean Prod 172:253–263. https://doi.org/10.1016/j.jclepro.2017.10.105
Feucht Y, Zander K (2018) Consumers' preferences for carbon labels and the underlying reasoning. A mixed methods approach in 6 European countries. J Clean Prod 178:740–748. https://doi.org/10.1016/j.jclepro.2017.12.236
Filippini R, De Noni I, Corsi S, Spigarolo R, Bocchi S (2018) Sustainable school food procurement: what factors do affect the introduction and the increase of organic food? Food Policy. https://doi.org/10.1016/j.foodpol.2018.03.011
Gadema Z, Oglethorpe D (2011) The use and usefulness of carbon labelling food: a policy perspective from a survey of UK supermarket shoppers. Food Policy 36:815–822. https://doi.org/10.1016/j.foodpol.2011.08.001
Garnett T, Mathewson S, Angelides P, Borthwick F (2015) Policies and actions to shift eating patterns: what works? A review of the evidence of the effectiveness of interventions aimed at shifting diets in more sustainable and healthy directions, Oxford
Gold S, Kunz N, Reiner G (2017) Sustainable global Agrifood supply chains: exploring the barriers. J Ind Ecol 21:249–260. https://doi.org/10.1111/jiec.12440
Grunert KG (2011) Sustainability in the food sector: a consumer behaviour perspective. 2:207–218
Grunert KG, Aachmann K (2016) Consumer reactions to the use of EU quality labels on food products: a review of the literature. Food Control 59:178–187. https://doi.org/10.1016/j.foodcont.2015.05.021
Grunert KG, Hieke S, Wills J (2014) Sustainability labels on food products: consumer motivation, understanding and use. Food Policy 44:177–189. https://doi.org/10.1016/j.foodpol.2013.12.001
Guenther M, Saunders CM, Tait PR (2012) Carbon labeling and consumer attitudes. Carbon Manag 3:445–455. https://doi.org/10.4155/cmt.12.50
Hallstein E, Villas-Boas SB (2013) Can household consumers save the wild fish? Lessons from a sustainable seafood advisory. J Environ Econ Manage. https://doi.org/10.1016/j.jeem.2013.01.003
Hartikainen H, Roininen T, Katajajuuri JM, Pulkkinen H (2014) Finnish consumer perceptions of carbon footprints and carbon labelling of food products. J Clean Prod 73:285–293. https://doi.org/10.1016/j.jclepro.2013.09.018
Hornibrook S, May C, Fearne A (2013) Sustainable development and the consumer: exploring the role of carbon labelling in retail supply chains. Bus Strateg Environ 24. https://doi.org/10.1111/cobi.12551
Hyams K, Fawcett T (2013) The ethics of carbon offsetting. Wiley Interdiscip Rev Clim Chang. https://doi.org/10.1002/wcc.207
ISO (2018) ISO 14024:2018 – Environmental labels and declarations -type I environmental labelling – principles and procedures
ISO (2016) ISO 14021: 2016 – Environmental labels and declarations – self-declared environmental claims (type II environmental labelling)
ISO (2010) ISO 14025:2010 -Environmental labels and declarations – Type III environmental declarations – Principles and procedures
ISO Central Secretariat (2012) Environmental labels and declarations. How ISO standards help. Environ. labels Declar. How ISO Stand. Help
Italian Ministry for the Environment Land and Sea (2011. Criteri ambientali minimi per il servizio di ristorazione collettiva e la fornitura di derrate alimentari [WWW Document]. https://www.

minambiente.it/sites/default/files/archivio/allegati/GPP/all.to_40_CAM_ristorazione_collettiva_e_derrate_alimentari_25.07.2011.pdf. Accessed 9.1.19

Iweala S, Spiller A, Meyerding S (2019) Buy good, feel good? The influence of the warm glow of giving on the evaluation of food items with ethical claims in the U.K. and Germany. J Clean Prod 215:315–328. https://doi.org/10.1016/j.jclepro.2018.12.266

Janßen D, Langen N (2017) The bunch of sustainability labels – do consumers differentiate? J Clean Prod 143:1233–1245. https://doi.org/10.1016/j.jclepro.2016.11.171

Klein A, Menrad K (2016) Climate-friendly food choices regarding fruit and vegetables: how German consumers perceive their competency and what supporting measures they would prefer. Acta Hortic 1132:183–190. https://doi.org/10.17660/ActaHortic.2016.1132.25

Kortelainen M, Raychaudhuri J, Roussillon B (2016) Effects of carbon reduction labels: evidence from scanner data. Econ Inq 54:1167–1187. https://doi.org/10.1111/ecin.12278

Lampert P, Menrad K, Emberger-Klein A (2017) Carbon information on vegetables: how does it affect the buying process? Int J Consum Stud 41:618–626. https://doi.org/10.1111/ijcs.12375

Lee SY (2008) Drivers for the participation of small and medium-sized suppliers in green supply chain initiatives. Supply Chain Manag. https://doi.org/10.1108/13598540810871235

Leire C, Thidell Å (2005) Product-related environmental information to guide consumer purchases – a review and analysis of research on perceptions, understanding and use among Nordic consumers. J Clean Prod. https://doi.org/10.1016/j.jclepro.2004.12.004

Liu T, Wang Q, Su B (2016) A review of carbon labeling: standards, implementation, and impact. Renew Sust Energ Rev 53:68–79. https://doi.org/10.1016/j.rser.2015.08.050

Lombardi GV, Berni R, Rocchi B (2017) Environmental friendly food. Choice experiment to assess consumer's attitude toward "climate neutral" milk: the role of communication. J Clean Prod 142:257–262. https://doi.org/10.1016/j.jclepro.2016.05.125

Maclnnis DJ, Jaworski BJ (1989) Information processing from advertisements: toward an integrative framework. J Mark. https://doi.org/10.2307/1251376

McGuire L, Beattie G (2016) Consumers and climate change: can the presence of others promote more sustainable consumer choice? Int J Environ Sustain 12. https://doi.org/10.4324/9781351277884-6

Meyerding SGH, Schaffmann AL, Lehberger M (2019) Consumer preferences for different designs of carbon footprint labelling on tomatoes in Germany-does design matter? Sustain. 11:1–30. https://doi.org/10.3390/su11061587

Monier-Dilhan S (2018) Food labels: consumer's information or consumer's confusion. OCL. https://doi.org/10.1051/ocl/2018009

Mungkung R, Sorakon K, Gheewala SH (2018) Ecolabelling and sustainable public procurement to promote sustainable consumption and production in Thailand. Chem Eng Trans. https://doi.org/10.3303/CET1863041

Nerlich B, Koteyko N (2009) Compounds, creativity and complexity in climate change communication: The case of "carbon indulgences". Glob Environ Chang. https://doi.org/10.1016/j.gloenvcha.2009.03.001

Neto B, Gama Caldas M (2018) The use of green criteria in the public procurement of food products and catering services: a review of EU schemes. Environ Dev Sustain 20:1905–1933. https://doi.org/10.1007/s10668-017-9992-y

Nie C, Zepeda L (2011) Lifestyle segmentation of US food shoppers to examine organic and local food consumption. Appetite. https://doi.org/10.1016/j.appet.2011.03.012

Onozaka Y, McFadden DT (2011) Does local labeling complement or compete with other sustainable labels? A conjoint analysis of direct and joint values for fresh produce claim. Am J Agric Econ 93:689–702. https://doi.org/10.1093/ajae/aar005

Onozaka Y, Nurse G, Thilmany McFadden D (2010) Local food consumers: how motivations and perceptions translate to buying behavior. Choices

Padel S, Zander K, Gossinger K (2010) Regional production and 'fairness' in organic farming: evidence from a CORE organic project. In: 9th European IFSA symposium, 4–7 July, Vienna (Austria)

Peattie K, Peattie S, Ponting C (2009) Climate change: a social and commercial marketing communications challenge. EuroMed J Bus 4:270–286. https://doi.org/10.1108/14502190910992693

Petty RE, Cacioppo JT (1986) The elaboration likelihood model of persuasion. Adv Exp Soc Psychol. https://doi.org/10.1016/S0065-2601(08)60214-2

Schaefer F, Blanke M (2014) Opportunities and challenges of carbon footprint, climate or CO2 labelling for horticultural products. Erwerbs-obstbau 56:73–80. https://doi.org/10.1007/s10341-014-0206-6

Sharp A, Wheeler M (2013) Reducing householders' grocery carbon emissions: carbon literacy and carbon label preferences. Australas Mark J 21:240–249. https://doi.org/10.1016/j.ausmj.2013.08.004

Sharp B (2010) How brands grow, what marketers Don't know. Oxford University Press

Sirieix L, Delanchy M, Remaud H, Zepeda L, Gurviez P (2013) Consumers' perceptions of individual and combined sustainable food labels: a UK pilot investigation. Int J Consum Stud 37:143–151. https://doi.org/10.1111/j.1470-6431.2012.01109.x

Smith J, Andersson G, Gourlay R, Karner S, Mikkelsen BE, Sonnino R, Barling D (2016) Balancing competing policy demands: the case of sustainable public sector food procurement. J Clean Prod 112:249–256. https://doi.org/10.1016/j.jclepro.2015.07.065

Soldi R (2018) Sustainable public procurement of food. https://doi.org/10.2863/1187

Stokes AM, Turri A (2015) Consumer perceptions of carbon labeling in print advertising: Hype or effective communication strategy? J Mark Commun 21:300–315. https://doi.org/10.1080/13527266.2012.762420

Styles D, Schoenberger H, Galvez-Martos JL (2012) Environmental improvement of product supply chains: proposed best practice techniques, quantitative indicators and benchmarks of excellence for retailers. J Environ Manag 110:135–150. https://doi.org/10.1016/j.jenvman.2012.05.021

Teisl MF (2003) What we may have is a failure to communicate: labeling environmentally certified forest products. For Sci

Teisl MF, Rubin J, Noblet CL (2008) Non-dirty dancing? Interactions between eco-labels and consumers. J Econ Psychol. https://doi.org/10.1016/j.joep.2007.04.002

Testa F, Iraldo F, Vaccari A, Ferrari E (2015) Why eco-labels can be effective marketing tools: evidence from a study on italian consumers. Bus Strateg Environ. https://doi.org/10.1002/bse.1821

Thøgersen J, Haugaard P, Olesen A (2010) Consumer responses to ecolabels. Eur J Mark 44:1787–1810

Thøgersen J, Nielsen KS (2016) A better carbon footprint label. J Clean Prod 125:86–94. https://doi.org/10.1016/j.jclepro.2016.03.098

Tsai WT (2017) Green public procurement and green-mark products strategies for mitigating greenhouse gas emissions—experience from Taiwan. Mitig Adapt Strateg Glob Chang 22:729–742. https://doi.org/10.1007/s11027-015-9695-3

Upham P, Bleda M (2009) Carbon labelling : public perceptions of the debate. Tyndall Cent. Manchester Sustain. Consum. Inst

Upham P, Dendler L, Bleda M (2011) Carbon labelling of grocery products: public perceptions and potential emissions reductions. J Clean Prod 19:348–355. https://doi.org/10.1016/j.jclepro.2010.05.014

Van Loo EJ, Caputo V, Nayga RM, Verbeke W (2014) Consumers' valuation of sustainability labels on meat. Food Policy 49:137–150. https://doi.org/10.1016/j.foodpol.2014.07.002

Vanclay JK, Shortiss J, Aulsebrook S, Gillespie AM, Howell BC, Johanni R, Maher MJ, Mitchell KM, Stewart MD, Yates J (2011) Customer response to carbon labelling of groceries. J Consum Policy 34:153–160. https://doi.org/10.1007/s10603-010-9140-7

Vecchio R, Annunziata A (2015) Willingness-to-pay for sustainability-labelled chocolate: an experimental auction approach. J Clean Prod 86:335–342. https://doi.org/10.1016/j.jclepro.2014.08.006

Vlaeminck P, Jiang T, Vranken L (2014) Food labeling and eco-friendly consumption: experimental evidence from a Belgian supermarket. Ecol Econ. https://doi.org/10.1016/j.ecolecon.2014.10.019

Wang HJ (2017) Determinants of consumers' purchase behaviour towards green brands. Serv Ind J 37:896–918. https://doi.org/10.1080/02642069.2017.1365140

Zepeda L, Sirieix L, Pizarro A, Corderre F, Rodier F (2013) A conceptual framework for analyzing consumers' food label preferences: an exploratory study of sustainability labels in France, Quebec, Spain and the US. Int J Consum Stud 37:605–616. https://doi.org/10.1111/ijcs.12041

Zhou S, Wang H, Li S, Chen Y, Wu J (2019) Carbon labels and "horizontal location effect": can carbon labels increase the choice of green product? Glob Ecol Conserv 18:e00609. https://doi.org/10.1016/j.gecco.2019.e00609

Conclusions and Recommendations

Climate change seriousness has been at the center of political and media attention for the last years. 2019, with the amazon forest fires and Australian bushfire awoke the consciences of many citizens and consumers. Greta Thunberg, recently elected person of the year by the Times, has guided a number of protests worldwide in which young people are demanding governments take action towards climate change. Moreover, data from the Eurobarometer confirms this concerns, in fact, they found that 93% of European citizens are concerned about climate change and 92% agree that the EU should take action to minimize emissions to reach climate neutrality in Europe by 2050 know (European Commission- Directorate General 2019). At the same time consumers are changing their attitude and behavior towards low carbon and carbon-neutral products. According to a Carbon Trust study of 2019 two-thirds of consumers support the idea of a recognizable carbon label to demonstrate that products have been made with a commitment to measuring and reducing their carbon footprint (Carbon Trust 2019). Moreover, 66% of consumers confirm they would feel more positive about companies that can demonstrate they are making efforts to reduce the carbon footprint of their products (Carbon Trust 2019). Furthermore, the Paris Agreement put at the center of the debate the offsetting mechanism and mitigation and adaptation strategies, institutionalizing Carbon neutrality as a suitable tool for the Low Carbon transition. Carbon markets discussion take the center also in the COP 25 in Madrid were governments tried to reach a consensus on how to regulate these emissions trading.

However, government failed to reach an agreement on Article 6 of the Paris Agreement, which details rules for carbon markets and other forms of international cooperation. Additionally, in Europe the new elected president of the European Commission Ursula von der Leyen made the European Green Deal the first point of her political agenda. The European Green Deal is a new growth strategy that aims to transform the EU into a fair and prosperous society, with a modern, resource-efficient and competitive economy where there are no net emissions of greenhouse gases in 2050 and where economic growth is decoupled from resource use.

A. Acampora et al., *Carbon Neutrality in the Agri-food Sector*,
https://doi.org/10.1007/978-3-030-88048-4

Moreover, the EU Commission announced that they will present the "Farm to Fork" Strategy in spring 2020 and launch a broad stakeholder debate covering all the stages of the food chain and paving the way to formulating a more sustainable food policy. In this context, companies are acting on these pressures implementing carbon-neutral strategies for their brands and products. A lot of case studies have been presented in this book showing how different are the approaches that companies are adopting. There is not a magic formula to reach the carbon neutrality that is universally valid. Companies are combining the tools available on the market according to their different goals, and their specificity. Companies are also experimenting with new approaches acting in a void of specific definitions and governance. The frontrunners of this transition are surely the companies of the agri-food sector. These companies have identified a specific competitive advantage and are exploiting it to become the pioneers and the reference model for the carbon neutral implementation. Indeed, agriculture has great potential in the fight against climate change, because it is the only sector that has the capacity to remove GHGs safely and cost-effectively from the atmosphere without reducing productivity. Moreover, the agricultural sector has the potential to benefit from synergies between climate change adaptation and mitigation measures. Improved agricultural practices can help mitigate climate change by reducing emissions from agriculture and other sources and by storing carbon in plant biomass and soils. Moreover, the sequestration of carbon in soils and vegetation can be used to offset greenhouse gas emissions. In sense, the industry can look to the agri-food sector to offset their emissions. Recently, we saw a number of agri-food companies committing to carbon neutrality goal. These companies are taking the lead of the low-carbon transition positing themselves as carbon neutral. Looking at the different carbon neutrality paths that agri-food companies are undertaking a major question arose in writing this book, is carbon neutrality a voluntary tool effective for the low-carbon transition?

Overall, we think this is a valuable tool for company that are genuinely trying to reduce their impact on the environment. The implementation of carbon neutral strategies by companies will accelerate the global reduction of GHG emissions and boost the investments in carbon offsetting projects that otherwise will not be implemented.

Developing a carbon neutral strategy can helps companies set long term strategy to reduce their carbon footprints and boost the transition towards a low-carbon agri-food sector. It represents a long-term umbrella project (emissions savings + energy savings + renewable energy + offsetting) that can help companies implementing short terms solutions (offsetting) while developing long-term carbon reduction projects. This is particularly important as carbon reduction programs are generally expensive and takes a lot of years before being operative so for companies sometimes it is difficult to see the advantages of these types of strategies. Furthermore, many companies will not easily adopt the carbon neutral process without relying on offsetting: in other words, offsetting is somehow comparable to a nudging strategy, which once get started forces to continue the carbon reduction process in the supply chain. Indeed, the implementation of a carbon neutrality strategy has been also as an activator of long-term supply-chain reduction programs. This is the case of the numerous insetting projects that we described in this book. Nevertheless, is important to note that to be truly sustainable the long-term strategy has to remain focused on reducing emissions as much as possible.

However, in this book we also pointed out as the carbon neutrality path is far from being perfect. Indeed, the recent interest for this process triggered the need to review this instrument in terms of governance and robustness. In particular, the carbon neutrality process should get out of the initial development phase and establish clearer rules for good governance if it wants to establish itself as a functional tool for the low-carbon transition.

We analyzed the different aspects that characterize the carbon neutrality strategy in the agri-food sector also highlighting shortcomings and difficulties specifically related to different products. From the analysis emerged as the governance system of the carbon neutrality still present some drawback that need to be overcome to assure the correct functioning of the system. Firstly, we highlighted as a common terminology surrounding the carbon neutrality theme is still missing and that the research field suffer from the lack of a common definition of carbon neutrality. We reviewed some of these definitions and we underlined as they don't differ just in formal aspects. Looking at the different terminology and related definition of the carbon neutrality, in fact, they present substantial differences (as for example CO_2 vs all GHG emissions, the use of the "unavoidable emissions" expression etc.). This aspect together with an unclear reference framework on how to achieve carbon neutrality, but more specifically on the tools available to companies to ensure and communicate carbon neutrality, have affected the development and diffusion of this process. More efforts should be done to clarify and simplify this path both for companies that want to implement it and for consumers who should be able to distinguish a "true" path and commitment to carbon neutrality from a greenwash path. In this book we have tried to clarify the most critical steps of the carbon neutrality process. However, some problems and critical aspects persist, above all in the agri-food sector, and much more can be done to improve this tool.

The most critical phase, also due to its more technical nature, is that of measuring the emissions associated with products and organizations namely the carbon footprint calculation.

In the measurement phase, even if as we saw in the third chapter of this book companies have a lot of tools to choose from, they all present the same shortcomings as they are mainly based on the same accounting methodology, the life cycle assessment (LCA). Particularly, these tools have scarce consideration of soil carbon sequestration effects and they are unable to precisely account for the use of pesticides, fertilizer and other chemicals. In fact, they are all based on LCA method that is currently imperfect and does not comprehensively assess some aspects that are critical for long-term sustainable food production (e.g. the emissions saved with climate smart agricultural practices, decreased soil quality and fertility, increased erosion, reduced ecosystem services due to intensification, biodiversity loss). Moreover, the LCA databases should be improved as the use of primary data when performing the LCA analysis. Moreover, these methodologies are currently unable to account for sustainable and climate mitigating practices applied in the agri-food sector. Nonetheless, even with its difficulties carbon footprinting is gaining a momentum in the sector as companies are developing climate-friendly strategies to take the lead in the low carbon transition and to do so the first step is to account for their associated emissions.

The second step is to reduce emissions. This is the most difficult phase for companies because it requires long-term investments and advanced technical skills to develop the corporate emissions reduction plan. This is true particularly for the agri-food sector as is difficult to implement sustainable practices and account for improvements in the farm managements strategies related for example to precision, regenerative and climate-smart agriculture. Moreover, are also missing guidelines for carbon reduction strategies for the sector. Here two orders of problems have been found: how to reduce emissions in the agri-food sector (e.g. which agricultural practices are better for climate) and how to measure the savings related to the implementation of these practices. In this sense, the French Carbon Agri project can be seen as a pioneer and a reference point in the sector in developing both guidelines to reduce emissions and methodologies to account for emissions reductions in the sector. This project is also dealing with the implementation and development of offsetting projects on the French territory in the agri-food sector. Indeed, offsetting projects are generally located in non- annex 1 countries because they were initially created mainly under the CDM mechanism This is particularly relevant as another shortcoming emerged from the analysis is the lack of offsetting project targeting specifically the agri-food sector. This is due to the fact, that, there isn't an international standard for developing carbon credits project in the sector. Neither, there are many offsetting projects in Annex I countries where with the guidance of SGDS these projects could boost sustainable development and be subject to more stringent controls from stakeholders. Another positive aspect of the French project, also from the point of view of the company marketing strategy, is the geographical closeness of company that are buying the carbon credits and the offsetting project that is generating them. This aspect is interesting because companies are more likely to be informed and involved in the offset project activities and this can create also a beneficial network between these two entities. This can help to overcome the mistrust related to the offsetting mechanism and to the carbon credits. These, in fact, have a mixed history filled with questions regarding additionality, permanence, fairness, and impact on broader decarbonization efforts. Given the complexity, controversy and uncertainties regarding the use of carbon credits to "offset" a company's footprint, another strategy that companies are exploring is developing carbon insetting practices, by investing directly in carbon reduction strategies in their own value chain. This approach can be beneficial for companies as in this way they secure the resilience of the chain to climate change also involving smallholder farmers and avoiding the risk of greenwashing related to purchasing carbon credits from offsetting projects.

Particularly, the analysis also showed as smallholder farmers are the most excluded from this carbon neutrality process. More efforts should be done to involve these operators and open up to offsetting possibilities. The little participation of as smallholder farmers is also related to another limitation emerged from the analysis that is linked to the fact that even if several offsetting certification schemes exist worldwide, it still misses an international standard for offsetting specific for the agri-food sector. The creation of such standard could expand the possibilities for the sector and enhance the role of the agri-food chain as source of solutions for climate change. Particularly, this standard should consider the potential role of offsetting projects in achieving different SDGS and assure the respect of human rights and sustainable development principles. Another aspect to consider is the lack of a common understanding and quantification methodologies to account for benefits and emissions reduction related to climate smart agronomics practices that has prevented smallholder farmers from participating in this process. If precise guidelines and calculation methods will be established, smallholder farmers may become key actors in the carbon neutrality process.

However, another aspect that clearly emerges from the analysis is that these strategies, being voluntary commitments of companies and are not related to any legislation or enforcement of specific laws, are mostly driven by marketing, social and economic reasons. Being outside of mandatory sphere, the carbon neutrality paths can take different directions, leaving to the companies the freedom to choose what fits better to their activity. In the book some case studies that shows the different paths have been presented. This is true for the four phases of carbon neutrality implementation (measurement, reduction, offsetting, documentation and validation) and for the use of the different tools available to companies (standard, verification, certification, accreditation and labeling). This variability in carbon neutrality paths can lead to companies to advertise their products and organizations as carbon neutral having behind different levels of robustness and transparency of carbon neutrality paths. In fact, another set of problems is related to how companies communicate their carbon neutral credential. Our analysis showed as this problem is related to both the ratio between reduction and offsetting and the use of self-declarations, second-party and third-party verification and certifications. Indeed, companies can advertise of being carbon neutral having a strategy that is more focused on reduction than offsetting, but they can also advertise of being carbon neutral just offsetting the 100% of their emissions. Another problem is related to the choice of scope of carbon neutrality. Companies can choose to work on scope 1 and 2 of their emission, having the most part of their emission related to scope 3, and advertise of being carbon neutral. These cases are possible because there is a lack of governance of this process, and mostly there is a lack of governance mechanisms for assuring the truthfulness and transparency of carbon neutral communication. Specificity, of the agri-food sector in this sense are the prevalence of the offsetting mechanism instead of the reduction strategy to reach the carbon neutral status. Particularly, this study analyzed the problem of greenwashing in the context of carbon neutrality but also in companies' sustainability reporting highlighting as these two are highly correlated. Greenwashing reputation of carbon neutrality claims is exacerbated by misleading information contained in companies' sustainability reports and environmental communication. Currently, companies carbon neutrality communication is very confusing and chaotic, and this can shed a negative light on the phenomenon. Often, looking to companies' websites it is very difficult to find complete and transparent information and this is aggravated by the lack of a common understanding of the framework of the carbon neutrality process. This lack of common understanding of companies, governments, and institutions is aggravating the greenwashing phenomenon that sometimes can be also involuntary (e.g. companies are not skilled enough to correctly communicate their carbon neutral credential or they are genuinely convinced that they are doing good). In this sense, would be recommendable to promote instruments for enhance transparency of carbon neutrality communication and provide guidelines on the process and how to report the different stage of the carbon neutrality process.

Other critical elements emerge when analyzing the robustness of the claims on carbon neutrality of the companies. First- second- and third-party verification and certification of companies' carbon neutrality path are not often specified. However, from the analysis emerged the third-party independent certification is little used by companies. So, to boost transparency and robustness of the carbon neutrality process use of the international standard and third-party verification and certification should be promoted. This could useful also to overcoming greenwashing issues and concerns from consumers through the establishment of a good carbon neutrality governance. In fact, there is still an offset mistrust related to the greenwashing issues and concerns from consumers that have to be overcome through a good governance (e.g. reporting rules, third party verification and certification etc.). Furthermore, consumers are still little informed about carbon neutral concepts, products, and labels so it is necessary to boost consumers awareness of carbon neutral concepts and labels trough a good institutional and industry communication. The history of food labels tells us that this type of communication works, and they are effective in guiding consumers towards more environmentally friendly choices and to inform and educate them about the environmental impact of the food's life cycle. However, carbon labels are overly complicated for consumers and a better design can help communicate better environmental and carbon information. Nevertheless, there are not many carbon neutral labels available to consumers. An institutional label for carbon information and carbon neutral products (e.g eco-label, EMAS etc.) is missing. The EU Commission is working on a European label with climate information (in the farm to fork strategy) and we think, if done properly, this would beneficial for the low carbon transition in the sector. In fact, credibility of carbon labels credentials, that in this case would be assured by the European institutions is a key factor to trigger consumers response towards these labels. In this sense, a certifying and auditing system that assures consumers about the truthfulness of labels criteria and information provided to them is essential to influence their decision-making process.

Indeed, to become an effective tool for the agri-food companies some issues related to the carbon neutrality process must be resolved. We summarized the major obstacles and issues to be solved, emerged during the redaction of this book, according to the different phases that have to be implemented by companies to reach carbon neutrality:

Measure

- Methodologies for accounting GHG emissions and removals have to be improved, especially for:
  - **Fertilizer**, **pesticides** and other **chemicals;**
  - **Mitigation agricultural** practices in LCA analysis;
  - **Soil carbon sequestration;**
  - Improve **LCA databases for agri-food supply chain** with regionalized and updated data;
  - Improve the use of **primary data.**

Reduce

- Guidelines for carbon management **reduction** strategies for the agri-food sector;
- Assess the potential of **precision agriculture** and **regenerative** agriculture.

Offsetting

- Promote **national offsetting program** (e.g. French model of Label Bas Carbone);
- Promote **sectoral reduction projects** (e.g. Aviation industry);
- Develop International **standards** for offsetting that are specific for the agri-food sector;
- Promote **additionality** of the carbon credits;
- Assure the respect of human rights and **sustainable development principles**.

Document and Validate

- Promote instruments for enhance **transparency** of carbon neutrality communication;
- Promote the use of the **international standard** and **third-party verification;**
- Clearer **definitions** and common **terminology;**
- Boost consumers awareness of carbon neutral concepts and labels;
- Overcoming greenwashing issues and concerns from consumers through a good governance;
- Use offsetting projects to promote **sustainable and regenerative agriculture** and different **SGDs**;
- Promote **capitals** and **funding** towards the agri-food sector to create new agro-forestry offsetting projects.

In our opinion for companies that wants to implement a carbon neutral strategy the tools to operate correctly in the markets are available, it is up to these companies to use it appropriately and to regulators to assure a good use of them and a good governance of the system. Surely, the private sector can do more, but it also needs more public intervention both in terms of rules of the game and also regulating and enforcing regulations on emissions. In this sense two examples stand out in this book the development of the France national offsetting program that will promote the double goal to offsetting companies' emission in the country and promote local environmental programs contributing to national reduction targets and sectoral emission reduction goals as in the case of the aviation sector that has been found effective in moving the sector to take action towards carbon neutrality.

Finally, the uptake of the carbon neutral strategy will benefit from a carrot and stick approach to fight climate change in the agri-food sector that can be summarized as follow:

- Stick 1: Emissions regulations
- Stick 2: Improving international governance on insetting/offsetting
- Stick 3: Regulating offsetting approach and ensuring sustainability plus avoiding double counting (INDC) (international coordination)
- Stick 4: Carbon taxing
- Carrot 1: Green procurement
- Carrot 2: Sovereign funds and green investment with blended finance
- Carrot 3: Facilitating access by farming communities to potential offsetting benefits of other firms (payment for environmental services driven by offsetting money)
- Carrot 4: Disseminating best available technologies and supporting know-how transfer to maximize insetting of agri-food chains
- Carrot 5: Financial support to certification and facilitating access to measurement, reduction, offsetting and reporting methodologies
- Carrot 6: consumer awareness – supporting consumer information disclosure and awareness of climate change and carbon neutrality

References

Carbon Trust (2019) Research reveals consumer demand for climate change labelling

European Commission- Directorate General (2019) Special Eurobarometer 490 "Climate Change" – Report. https://doi.org/10.2834/00469

Index

A
Accreditation systems, 106, 108
Adaptation measures, 34
Additionality, 65, 149, 180, 183
Advertising, 156, 157
Afforestation, 37, 46, 77, 78, 148, 149
AFOLU emissions, 37
Agricultural activities, 30, 147
Agricultural practices, 31, 59, 178–180, 183
Agriculture, 29–41, 44, 45, 61, 77, 82, 120, 127, 145, 148, 149, 178, 180, 183
Agri-food, 1, 29–30, 82, 108, 115–136, 143–145, 155, 170, 178, 181–184
Agri-food businesses, 126
Agri-food sector, 29–47, 51, 58–62, 115, 126, 178–181, 183, 184
Agro-ecological practices, 44
Agroforestry, 37, 46, 78, 87
Alter Eco, 74
Anaerobic digestion, 118
Animal breeding, 118
Animal pasture, 87
Animal welfare, 159, 171
Anthropogenic emissions, 7, 20

B
Barilla Center for Food and Nutrition, 6, 52, 121–126
Barriers, 37, 64, 139, 140, 145–148, 157, 167–169
Baseline scenario, 149
Behavioral change, 140, 156, 167
Ben & Jerry's, xiii, 87
Biennial reports (BRs), 14
Biennial transparency report (BTR), 14, 16
Biennial update reports (BURs), 14–16
Biodigesters, 128
Biodiversity, 29, 58–60, 78–80, 91, 139, 179
Bio-energy, 119
Bioenergy with Carbon Capture and Storage (BECCS), 37
Biogas, 46, 119
Biomass, 33, 118, 119, 178
Bonn challenge, 45
Broader scope insetting, 85

C
CAP2'ER, 150
Carbon accounting, 21, 51, 53, 79
Carbon Action initiative, 68
CARBON AGRI, 148–150, 180
Carbon budgets, 10, 11, 23
Carbon Capture and Storage (CCS), 8
Carbon credits, 45, 46, 65, 71, 76, 85, 87, 92, 95, 101, 107, 108, 131, 149, 169, 180, 183
Carbon cycle, 23
Carbon dioxide (CO_2), 4, 20–23, 30, 46, 81, 112
Carbon dioxide equivalent (CO_2e), 21, 76, 122
Carbon dioxide removal (CDR), 36, 37
Carbon disclosure, 63
Carbon disclosure initiatives, 63, 68–70
Carbon disclosure project (CDP), 53, 54, 68, 69, 95
Carbon equivalent, 19

A. Acampora et al., *Carbon Neutrality in the Agri-food Sector*,
https://doi.org/10.1007/978-3-030-88048-4

Carbon footprint (CF), 2, 20, 51–56, 58–62, 65, 71–75, 82, 83, 87, 90–93, 95, 96, 98, 100, 106–108, 111, 121–123, 125–131, 133, 140, 148, 157, 160, 170, 171, 177, 179
Carbon-free, 65, 161, 169
Carbon insetting, 82–88, 180
Carbon intensity, 24, 95, 119
Carbon label information, 142
Carbon labelling, 119, 148, 155–158, 164
Carbon labels, 73, 109, 119, 120, 141, 142, 155, 157–165, 177, 182
Carbon leakage, 147
CarbonManagement Plan, 101
Carbon markets, 71, 72, 76, 77, 81, 85, 147, 177
Carbon measurement, 72, 73, 90–92, 95, 96, 98, 100, 106, 130
Carbon negative, 22
Carbon neutral certifications, 99, 102, 103, 109, 129, 131
Carbon neutral country, 127
Carbon neutrality, 1–25, 56, 63, 65, 71–108, 111, 115–136, 139, 148–150, 161, 164, 169, 177–184
Carbon neutrality claims, 75, 101, 169, 181
Carbon neutrality commitment, 75
Carbon neutrality standards, 54
Carbon neutrality statement, 75
Carbon neutrality strategies, 88, 93, 106, 131, 132, 148, 178, 179
Carbon neutral labels, 72, 75, 103, 107–108, 142, 155, 157, 164, 169–172, 182
Carbon neutral products, 142, 155, 169, 177, 182
Carbon Offset and Reduction Scheme for International Aviation (CORSIA), 150, 151
Carbon offsetting, 64, 65, 82, 92, 129, 169
Carbon offsetting projects, 45, 178
Carbon positive, 22, 72, 74–75
Carbon protocol, 144
Carbon quantification, 72, 73
Carbon reductions, 65, 71–73, 76, 78, 79, 88, 101, 109, 111, 128, 144, 148, 164, 165, 169–172, 178, 180
Carbon-related regulations, 144
Carbon sequestration, 36, 37, 43, 59, 92, 121, 131, 179, 183
Carbon sinks, 8, 20, 95
Carbon standards, 47, 95, 101, 148
Carbon stocks, 35, 36, 118
Casino, 73
Casino Carbon Index, 73
Certifications, 45, 71, 72, 79, 86, 90–96, 100, 101, 103–109, 118–120, 128–131, 145, 155, 161, 166, 181, 182, 184
  bodies, 86, 105, 107, 156
  label, 105
  programs, 105
Certified Emission Reductions (CERs), 78, 80
Chocolats Halba, 87, 88
Circular economy, 8
Clean Development Mechanism (CDM), 77–80, 92, 94, 101, 180
Climate Active Carbon Neutral Standard, 102–103
Climate Active certification, 102
Climate change, 1–16, 29–42, 44, 45, 68, 69, 79, 82, 85, 87, 109, 115, 119, 120, 122, 140–142, 144, 145, 147, 155, 158–161, 165–169, 177, 178, 180, 181, 184
Climate change adaptation, 32, 178
Climate change mitigation, 8, 12, 13, 33, 42, 63, 158, 166
Climate, Community and Biodiversity Standard (CCBS), 79
Climate consciousness choices, 166
Climate Disclosure Standards Board (CDSB), 69
Climate-friendly, 78, 115, 117, 141–147, 159–161, 167, 168, 170, 180
Climate-friendly agricultural practices, 145
Climate-friendly diets, 140
Climate friendly technology, 119
Climate negative, 23
Climate neutral, 3, 5, 7, 18, 19, 65, 87, 88, 109, 111, 112, 141, 142, 169
Climate neutrality, 19, 21, 22, 88, 177
Climate neutral products, 29, 141–142
Climate positive, 22, 74
Climate resilience, 81
CO_2 emissions, 8, 18, 19, 21, 88, 128
Co-benefits, 37, 39, 46, 78, 80, 149
Compensation, 88, 148
Compensation projects, 106, 108
Compliance markets, 76
Composting, 59, 90, 118
Consumer awareness, 1, 120, 156, 184
Consumers, 1, 6, 36, 51, 63, 64, 66, 73, 103, 105, 115, 118–120, 126, 129, 131, 133, 139–142, 144, 147, 155–170, 177, 179, 182–184
  perceptions, 155–158
  preferences, 126, 160, 164, 166, 167
Control system, 108, 112

Coopedota, 129–131
Coppicing, 148, 149
Corporate carbon norms, 143
Corporate emissions, 88, 180
Corporate risks, 144
Corporate social responsibility (CSR), 65, 66, 76, 117
Cradle-to-gate, 55, 130, 131, 133
Cradle-to-grave, 51, 55, 130, 131
Crop breeding, 118
Croplands, 37
Crop production, 30

D
Decarbonization, 6, 8, 10, 19, 22, 69, 71–72, 146, 180
Deforestation, 36, 37, 43, 44, 47, 68, 74, 78, 106, 118
Demand-side measures, 35
Desertification, 36, 37, 39, 40
Developing countries, 4, 15, 16, 29, 42, 77, 78, 148
Dietary changes, 37, 120
Dilmah, 94
Direct emissions, 53, 93, 121
Disease management, 118
Distributors, 97, 133, 156
Dole, 127–128
Double-counting, 65, 184
Double Pyramids, 121, 122
Drop the C' program, 73

E
Eco-friendly behavior, 160
Eco-labelling, 119
Ecological Footprint, 122
Ecosystem conservation, 37
Ecosystem services projects, 78
Eligibility criteria, 149
Emission factors, 13
Emission inventory, 59
Emissions reductions, 11, 13, 45, 69, 71, 76, 81, 86, 87, 95, 133, 144, 149, 150, 180, 181
Emissions reduction strategy, 133
Emissions reduction targets, 9
End-of-life, 61
Energy consumption, 2, 41, 140, 144
Energy efficiency, 7, 41, 68, 74, 94, 95, 119
Energy efficiency projects, 46, 106
Energy-efficient cookstoves, 45
Enhanced Transparency Framework (ETF), 12–16
Enteric fermentation, 30, 121
Enteric methane, 133
Environmental concerns, 2, 140, 142, 160
Environmental crisis, 31
Environmental damage, 29, 61
Environmental problems, 1, 140, 159
Environmental Product Declaration (EPD), 106, 107, 156
Environmental risks, 144
Environmental standards, 120
European Union's Emissions Trading Scheme (ETS), 76
EXPO, 92
Extreme weather events, 31

F
Facilitative sharing of views (FSV), 15
Fair Trade, 129, 142, 160
Fair Trade labels, 160
Fairtrade Climate Standard, 45, 46
Fairtrade International, 45
Farmers, 31, 46, 60, 74, 87, 88, 115, 118, 119, 129, 145–147
Farming activities, 118
Farm management data, 129
Farm practices, 35, 118
Farm to Fork, 125, 178, 182
Feed optimization, 118
Fertilizers, 30, 33, 58, 60, 61, 82, 117, 118, 127, 130, 160, 179, 183
First-party audit, 104
First-party certification, 105
Fisheries, 30, 140
Five Founders, 131–135
Fluorinated gases, 19
Food and Agriculture Organization of the United Nations (FAO), 33, 42, 117, 119, 139
Food companies, 115, 119, 158
Food manufacturer, 87
Food processing, 30
Food processor, 119
Food production, 29, 31, 61, 121, 140, 156, 158
Food production systems, 139
Food products categories, 163
Food safety, 31
Food security, 36, 37, 39, 40
Food's life cycle, 155, 182

Food storage, 119
Food supply chains, 30, 36, 116
Food value chains, 33
Food waste, 29, 36, 120, 171
Forest biomass, 36
Forest carbon credits, 47
Forest degradation, 37, 43
Forest landscape restoration (FLR), 45
Forest Trends' Ecosystem Marketplace, 81–84, 151
Forestry, 29, 30, 36, 40, 45, 46, 78, 79, 81, 128, 148, 149
Forestry projects, 46
Fossil fuels, 23, 41
4 per 1000 initiative, 44
Free range, 160
French Label Bas Carbone, 148–150
Freshwater withdrawals, 29
Fuel consumption, 61, 127
Functional Units, 60, 65, 133, 169

G

GHG inventories, 15, 16, 57
GHG Protocol, 52, 54–59, 74, 90, 93, 98, 100, 108
GHG Protocol Corporate Standard, 52, 53, 58, 70
GHG protocol for guidance on carbon removals and land use, 54–57
GHG Protocol Product Lifecycle Accounting and reporting standard, 53
GHG Protocol Value Chain Standard, 53
Global anthropogenic emissions, 29
Global Reporting Initiative (GRI), 65, 66
Global Reporting Initiative (GRI) Standards, 65
Global warming, 1, 5, 7, 21, 31, 32, 47, 51, 59, 71
Global warming potentials (GWP), 19, 37, 59
Gold Standard 45, 46, 77, 80, 88, 92, 95, 98
Gran Cereale, 106, 107
Grasslands, 37, 82, 118, 150
Greenhouse gas (GHG) emissions, 3–8, 20, 21, 30, 42, 55, 72, 76, 78, 81, 91, 95, 118, 122, 128, 162, 178
Greenhouse gas inventories, 54
Green labels, 142
Green procurement, 120, 171, 184
Green Public Procurements (GPP), 144, 155, 170–172
Greenwashing, 63–68, 129, 155, 166, 169, 180–183

H

Heat tolerant crops, 117
Heineken, 73
Hotspots, 51, 61, 91, 144
Hydrogen, 21

I

Indirect emissions, 52, 53, 58, 92, 93, 100, 120, 171
Information asymmetry, 156
Initial baseline level, 76
Input companies, 115, 117
Input-output analysis, 133
Insetting, 71, 82, 83, 85–89, 98, 108, 184
Insetting Programs, 86
Insetting Program Standard (IPS), 86, 87, 98
Insetting Projects, 86, 179
Intended Nationally Determined Contributions (INDCs), 8, 12
Intergovernmental Panel on Climate Change (IPCC), 4, 7, 20, 22, 23, 32, 36, 37, 40, 41, 79, 147
International Carbon Reduction and Offset Alliance (ICROA), 76, 85, 92
International Civil Aviation Organization (ICAO), 150
International Emissions Trading Association (IETA), 76, 77
International Organization of Standardization (ISO), 46, 52, 53, 55, 56, 90–92, 96, 105, 108, 127, 155, 156
International Platform for Insetting (IPI), 85–87
International Standards, 52, 54, 97, 98, 105, 107, 108, 129, 180–183
Inventory data, 60, 61
ISO 14021
  2016, 156
ISO 14024, 156
ISO 14025
  2010, 156
ISO 14064-1, 52, 55, 98, 100, 107
ISO 14067, 53, 55, 98
ISO 9001, 129

J

Joint Implementation, 101

K

Kyoto Protocol, 5, 23, 76, 78

L

Labelling, 61, 97, 160, 161
Labelling schemes, 141
Land degradation, 36, 37, 39, 40
Land productivity, 29
Land restoration, 37
Landscapes, 45, 58, 59, 87, 118
Land tenure, 146
Land-use change emissions, 23
La Poste Group, 92
LCA methodology, 55, 106, 125, 133
Leakage, 147
Life Cycle Assessments (LCA), 51, 54, 56, 58–61, 71, 100, 122, 133, 134, 156, 169, 179, 183
Livestock, 30, 37, 121, 125, 133
  manures, 30
  methane, 82
  production, 121
Local communities, 6, 11, 87, 88
Local foods, 160
Long-term low greenhouse gas emission development strategies (LT-LEDS), 8–12
Low carbon choices, 139, 155, 158, 161
Low carbon economy, 21, 69, 115
Low carbon food products, 157
Low carbon neutral food products, 140
Low carbon procurement (LCP), 171
Low carbon products, 120, 141, 144, 160, 170
Low carbon transition, 71–113, 120, 143, 177–180, 182
Low carbon transformation, 22
Low emissions transformation, 22
Low yielding systems, 121

M

Manures, 118, 121, 150
Market drivers, 139, 143–145, 148
Marks & Spencer, 95
Max Burgers, xv, 74
Measurement, reporting and verification (MRV) framework, 12–16
Mediterranean Diet, 121
Methane (CH_4), 10, 19, 21, 30, 35, 133
Mitigation actions, 13, 43, 54
Mitigation measures, 33, 35, 36, 178
Monogastric animals, 119, 121
Multilateral assessment (MA), 15

N

National Communications (NC), 16
Nationally Determined Contributions (NDCs), 8–14, 33–35
Nature-based solutions, 81
Negative hotspots, 127
Nespresso, 86, 87
Net zero carbon emissions, 10, 18, 20, 22
Net zero commitments, 16, 18
Net zero emissions, 16–20, 102, 145
Net zero GHG emissions, 19, 22
Nitrogen-fixing legumes, 118
Nitrous oxide (N_2O), 19, 21, 23, 30, 35
Non-CO_2 emissions, 19, 35, 41
Non-permanence risks, 149
Nutrient use, 118

O

Offset certification, 108
Offset strategy, 90–96, 101, 106, 133, 136
Offsetting projects, 24, 56, 71, 180, 181, 183
Organic farming, 119
Organizational carbon footprint (OCF), 51–53, 55, 59, 171
Overconsumption, 120
Ozone, 21

P

Packaging, 30, 51, 62, 73, 85, 90, 97, 112, 119, 123, 125, 126, 133, 156, 163, 171
Paddy rice, 30
Paris agreement, 4–9, 14, 15, 21, 34, 143, 177
PAS 2050, 53, 55, 98, 100, 127, 148
Pesticides, 58, 60, 61, 160, 179, 183
Pest management, 118
Plant-based foods, 121
Plant-for-the-Planet, 44, 45
Plan Vivo, 78–80, 85–87, 112
Plan Vivo Certificates, 78, 80, 86
Point of sale, 120, 170
Premium price, 141, 142, 144
Private driven standard, 108
Private standards, 91, 97, 98, 105, 107
Product carbon footprint (PCF), 51, 53–55, 157
Project Gigaton, 73
Public driven standard, 108

Publicly Available Specification PAS 2060, 99
PUR Projet, 74, 86, 87
Purchase intentions, 142, 163

Q

Qualifying Explanatory Statement (QES), 101

R

Rainforest Alliance, 46–47, 129
Raw materials, 51–53, 73, 123
Recycling, 2, 51, 73, 92, 119
Reduction strategies, 90–92, 95, 96, 106, 180, 181, 183
Reforestation, 37, 45, 46, 77, 78, 88, 95, 118
Regenerative vegetation offset projects, 133
Remaining emissions, 3, 101, 102
Renewable energies, 4, 9, 10, 33, 41, 73, 74, 77, 90, 94, 119, 178
Renewable energy projects, 46, 95, 133
Resource use, 29, 177
Retail, 30, 51, 88, 115, 118, 120, 170
Retailers, 53, 73, 115, 120, 144, 155, 161, 170–172
Ruminants, 119, 121

S

Science based targets (SBTs), 69, 92
Science Based Targets initiative (SBTi), 69, 70, 95
Scope 1 emissions, 53, 58, 65, 70, 91, 93, 95, 100, 181
Scope 2 emissions, 58, 70, 100
Scope 3 emissions, 53, 58, 100
Second-party audit, 104
Second-party certification, 105
Sky, 94, 95
Smallholder farmers, 78, 82, 180, 181
SOCIALCARBON, 78–80
Soil biodiversity, 60
Soil carbon sequestration (SCS), 36
Soil fertility, 60
Soil hydrology balance, 60
Soil organic carbon (SOC), 44
Soil organic carbon management, 37
Soil structure, 60
Solar power, 133
Stakeholder involvement, 46
Supermarkets, 157
Suppliers, 73, 83, 85, 95, 104, 118–120, 139, 144–148, 170–172
Supply chains, 30, 51, 71, 82, 85, 87, 88, 95, 115–120, 133, 144, 145, 155, 171, 178, 183
Supply-side measures, 35
Sustainability labels, 160
Sustainability reporting, 63, 65–68, 181
Sustainable agricultural practices, 32, 129, 146
Sustainable development, 7, 13, 37, 45, 77, 78, 88, 97, 98, 180, 181, 183
Sustainable diets, 139, 140
Sustainable food production, 37, 58, 179
Sustainable forest management, 37, 47
Sustainable land management, 36
Sustainable procurements, 119, 144

T

Task force on Climate-related Financial Disclosure (TCFD), 69
Tesco, 73, 161, 162, 170
Third-party audit, 104
Third-party certification, 90, 105
Third-party certified, 24, 101, 107
Traders, 53, 115, 119
Transport, 22, 30, 51, 60, 73, 90, 92, 95, 106, 119, 121, 125, 140
Tree-planting initiatives, 44–45

U

Unilever, 74
United Nations Collaborative Programme on Reducing Emissions from Deforestation and forest Degradation (REDD+), 42
United Nations Development Programme (UNDP), 42
United Nations Environment Programme (UNEP), 6, 20, 24, 42
United Nations Framework Convention on Climate Change (UNFCCC), 5–7, 9, 12–16, 21, 33, 77, 78, 80
United Nations Global Compact, 69
Upstream carbon insetting, 83

V

Validations, 46, 52, 55, 71, 100, 101, 181
Vegan, 120, 140
Vegetarian, 120, 140
Verifications, 13, 14, 46, 52, 53, 55, 71, 72, 77, 90, 91, 103–112, 156, 181–183
Verified Carbon Standard (VCS), 77, 79, 80, 90, 92, 95, 98, 106, 107, 133

Verified Carbon Units (VCUs), 77, 80
Verra, 77, 79, 80
Voluntary offset markets, 76, 81–82
Voluntary standards, 77, 96–103

W
Walmart, 73
Warehouses, 73, 133
Wasa, 90
Waste management, 115, 171
Waste management projects, 46
Water Footprint, 122, 125
Water scarcity, 32
Wholesaler, 133
Willingness to pay (WTP), 139, 141–142, 161
World Business Council for Sustainable Development (WBCSD), 54–56, 59, 100
World Resources Institute (WRI), 9, 22, 54–56, 59, 69, 100
Worldwide Fund for Nature (WWF), 69

Y
Yields, 58, 60, 87, 118, 121

The manufacturer's authorised representative in the EU is Springer Nature Customer Service Centre GmbH, Europaplatz 3, 69115 Heidelberg, Germany. If you have any concerns regarding our products, please contact ProductSafety@springernature.com

Printed and bound by CPI Group (UK) Ltd, Croydon, CR0 4YY
22/09/2026
02220946-0001